WHY DARKNESS MATTERS
The Power of Melanin
In The Brain

Authored by
Edward Bruce Bynum
Ann C. Brown
Richard D. King
T. Owens Moore

PRINTING & PUBLISHING

KeithRyales.com

Front cover Illustration by Keith Ryales

Copyright© 2022 by Edward Bruce Bynum, Ph.D.

All rights reserved.

Third Edition, First Printing

No part of this book may be reproduced, stored in retrieval systems or transmitted in any form, by any means, including mechanical, electronic, photocopying, recording or otherwise, without prior permission of the publisher.

ISBN#: 978-1-884897-05-4

TABLE OF CONTENTS

DEDICATION iv

THE NEUROMELANIN HYPOTHESIS v

CHAPTER 1 1
Neuromelanin: A Highly Sensitized Sensory-Motor Network
T. Owens Moore, Ph.D.

CHAPTER 2 35
Neuromelanin: What Is Its Importance in Neural Tissue?
Ann C. Brown, Ph.D.

CHAPTER 3 73
The Clinical Use of Bliss: A Standardized Technique for Conscious Intervention Into the Functioning of the Autonomic Nervous System *Edward Bruce Bynum, Ph.D.*

CHAPTER 4 119
Neuromelanin: A Black Gate Threshold; The I-33 Tissue Of Heru, Historical, Neurophysiological, And Clinical Psychological Issues *Richard D. King, M.D.*

CHAPTER 5 161
Fire Atoms Ignite Inner Vision *T. Owens Moore, Ph.D.*

CHAPTER 6 173
The Farther Reaches of Eldership and the Dreamlife of Families *Edward Bruce Bynum, Ph.D., ABPP.*

EPILOGUE 216

POSTSCRIPT 217

GLOSSARY 223

In Memoriam: The Passing of a Giant in the Mental Health Field *T. Owens Moore, Ph.D.* 227

AUTOBIOGRAPHICAL STATEMENTS 230

DEDICATION

To our families

whose Love and beauty sustain us,

To our ancestors,

whose courage and vision

Has shown us the way,

To our progeny

Whose lives dream and arch into the future;

To all the hues and faces of humanity

With whom we share this inner

luminosity and journey.

THE NEUROMELANIN HYPOTHESIS

"All traces of psychological subjects, including 'psychotherapy,' were practiced in Africa by the Egyptians and long pre-dated the Greek, Roman, and Hebrew tradition in which much modern Western psychology is rooted. Imhotep... [is] the first figure of a physician to stand out clearly from the midst of antiquity."

-A. K. Tay, "Psychology in Africa," *The Unesco Courier*

The thrust of this book, *Why Darkness Matters,* on the reality of melanin in the brain or the neuromelanin hypothesis is to acquaint a larger audience with the history, complexity, and importance of this vital and at times controversial area of science, psychology, and neuroscience. A searching review of its contents, however, will quickly reveal that it touches upon a number of sensitive areas in the living fabric of our society and, by historical extension, the dynamics of many other contemporary cultures and sociopolitical worlds.

The importance of variable surface skin melanin in Western and European-influenced societies and the subsequent psychodynamics of racism and color are well known and documented (Akbar, 1984; Fanon, 1967; Welsing, 1991). Indeed, this aspect of the phenomenon has been the source of not only almost endless academic debates but also volatile political and social movements. No one who is only vaguely aware of the confluence of race and ethnic realities in the United States alone cannot fail to see its impact and current expressions. Its dynamics were deeply woven into the intimate social and economic fabric of the republic from the earliest days of its existence (Bennett, 1966; Stampp, 1956).

We are almost fixated on race and its cyclical eruption in the streets, classrooms, and courts of America. We have had great difficulty seeing beyond it. The attempts to become a "color blind" society, while noble in impulse, have only painfully avoided certain issues, made others worse, and by fixation on mere surface skin dynamics, made us blind to other, perhaps deeper, more luminous realities. This volume will go in an altogether different direction and offer a new paradigm from which to observe and experience this phenomenon.

The seminal influence of inner melanin, or brain melanin, in human consciousness is not as well known as surface, or skin melanin and yet implicates all human beings without regard to so called "race" or ethnic diversity. This new wave of academic and clinical interest in melanin is actually the second wave, the first occurring in the 1980s. Each wave takes us further and deeper into the ocean of this area of science. The first wave sometimes hits upon the reefs of political polemics due to certain exaggerated claims made for it and, at times, the lack of clear scientific and clinical depth in its presentation to the public. It was a risky and controversial area to study, given the climate at that time, a climate that persists to some degree even today. However, it must also be emphasized here that certain problems arose in its study because of the **independent conceptual vision it unfolded for African peoples with the enormous potential and implications it has for science and society.**

You see, it was not so much the "objective" data that was in question but rather the psychohistorically informed scientific lens through which the data was focused that was crucial. It suggested neither a psychodynamic nor cognitive behavioral view of mind and consciousness, but rather a scientific vision that was not tacitly even within the sphere of Eurocentric intellectual thought. It harkened back to older, more expansive paradigms of mind, life, and science that illuminated and held sway over the ancient world for millennia. It also opened a pathway out of the obsessive preoccupation with the painful but addictively familiar framing of issues of race and color, around which many individuals and even institutions were and are focused. These powerful interests made for a difficult launch in a new direction. Hopefully, this new wave can steer clear of these reefs and stay true to the potential it carries for both science and African American studies in particular. This book makes its contribution in the area of neuroscience.

Neuroscience was actually born from the womb of Kemetic Egyptian neuroanatomy. By the 18th Dynasty there were thousands of medical papyri used in the medical "houses of life" or per ankh, which became the model for later Greek medical practice (Ghalioungui, 1973; Breasted, 1984). Indeed, due to the necessity for both battlefield trauma medicine and the elaborate process of mummification, a great deal was known about human anatomy (Finch, 1990). Sadly enough, of the thousands of medical texts written, only 10 have come down to us, of which the Edwin Smith and Ebbers papyri are the most well known.

The Edwin Smith "medical" papyrus is actually a surgical treatise and a copy of a much older papyrus dating back centuries. It covers only the head and neck. The other parts of the text, unfortunately, are lost to history. It details some 50 specific anatomical sites in the face, neck, and cranium alone. The Kemetic Egyptians were well aware of the cerebral gyri and the meninges of the brain, the dura mater, pia mater, and arachnoid, along with certain injury-related behavioral expressions in clinical practice. They were also aware of the crucial importance of the cerebrospinal fluid that bathes the brain, the inner mid-brain regions, and the spinal cord (Finch, 1990). Kemetic medical anatomy identified well over 200 clinical sites, of which almost 100 are located in this surviving text. Even after the geopolitical and military fall of Kemet, it remained the graduate school of the ancient Western World for he Greeks and Romans. As Homer said in the Odyssey, "In medical knowledge, Egypt leaves the rest of the world behind." It was not until the 19th century that the amount and detail of such medical knowledge was surpassed in Europe. In this sense neuroscience has been a living part of medicine and psychology from the earliest eras of African civilization and underlies much of contemporary medical practice.

Traditionally, Western science has adopted two divergent stances in regard to these two branches of African medical and psychological science. Either it has completely ignored it despite its existence, as evidenced in the above data on the African background of medical science in both internal medicine and the fundamentals of neuroscience, both of which are rarely if ever mentioned in our standard medical and scientific texts. Indeed, the unconscious was actually discovered in ancient Egypt and given the names of the Primeval Waters of Nun and the Amenta (Hourning, 1986; King, 1990). Carl Jung's concept of the archetype derives from the north African scholar Saint Augustine's *Principales,* as well as Jung's notions of anima and animus, which came from the anthropologist Sir Edward Tylor's "study" of Africans in his *Primitive Culture* (1871). And the Greek word "psyche" itself is a derivation of the older Egyptian words "Khe" for soul and "Su" for she, Su-Khe (Massey, 1881). Such facts are either ignored or else the knowledge is greatly distorted and held up to ridicule as simply "primitive" superstition or the "dark sciences" when it comes to areas of spiritual practice and certain anomalous experiences. Popular movies and other cultural motifs have a rich history of showing Africans practicing "black magic," which the West is both amused by and secretly terrified of for many reasons that go beyond the present scope of this book.

The chapters here are organized around the subject of brain melanin, or neuromelanin, putting the material in a new light. Melanin as a social and skin perception experience is a complex social, political, and cultural phenomenon. Brain, or neuromelanin, however, is an altogether different phenomenon that has less to do with various human prejudices and fears and more to do with human nervous system functioning, evolutionary unfoldment, and, ultimately, as we shall suggest, consciousness itself. These chapters will unfold sequentially and present in both academic and medical clinical clarity that melanin, and especially neuromelanin, with its affinity for bio conductivity and luminosity, is present in significant amounts in crucial areas of the cerebral and central nervous systems of *all* human beings. This has medical, psychological, and even spiritual implications.

But what does this mean? Why is it that melanin appears to be an "organizing molecule" in the early embryological and structural unfoldment of higher-level mammals and primates (Barr, 1983)? Is this just a biological artifact or does it imply something deeper about human experience and the process of evolution itself? What does this warm, living dark matter have to do, if anything at all, with the ubiquitous cold dark matter of the cosmos? Historically, African epistemological paradigms have stressed that the macrocosm reflects the microcosm through the principle of correspondence: "as above, so below," "as within, so without." This is the great organizing vision of the primogenitor Tehuti, or Thoth, who the Greeks would later adopt and call Hermes Mercurius Trismegistus (Chandler, 1999; Copenhaver, 1992; Kybalion, 1940). In contemporary scientific language we would refer to certain aspects of this vision in neuroscience and quantum mechanics as the holonomic paradigm (Pribram, 1991; Bohm, 1980).

About the brain itself only a small shore of an almost infinite ocean is presently known. We know that the adult human brain weighs roughly three pounds, has a thick jelly-like consistency, and is about 90 percent the content of seawater. Its surface looks somewhat like the bark of a tree, from which it derives its Latin name, "cortex." It has two hemispheres and four distinct regions, or lobes: the frontal, occipital, parietal, and temporal. Each lobe has localized functions and all work in intimate biochemical, bioelectric, and resonate coordination with each other. The cortex is that part of the brain that is most developed in modem humans. This cortex is a folded sheet of neurons, with 99 percent of its thickness between 1 and 4.5 mm. The average thickness, which is associated with increased "activity," is about 2.5 mm with regional variations. Cortical thinning, on the other hand, is due generally to disease and/or aging and is often specific to regions of the brain rather than global.

The brain is a very versatile and adaptive organ with many redundant features. Research in the neurosciences is always producing fascinating surprises for both the scientist and the general public. Brain imaging devices such as the MRI have revealed that the brains of men and women are different. Men's brains are on average 10 percent larger than women's while women generally demonstrate more functional connections between the two hemispheres of the brain at the connecting region of the corpus callosum. In certain regions of the brain women show a greater density of neurons and use more areas when solving a problem. Men, on the other hand, tend to be more focused in specific regions of the brain when working on a cognitive problem. None of this implies any innate or ultimate superiority or inferiority between men and women, only diversity and difference related to specific tasks, situations, and conditions. There is even evidence indicating that learning new skills actually increases neuromelanin concentration in certain regions of this marvelous organ and fountainhead of human intelligence and creativity. This last finding is a most intriguing.

Despite being hidden from the sun, the cortex is dark or "gray" in color because it is rich in neuromelanin, or brain melanin, and the darker and richer this neuromelanin has become through evolution the more cognitively sophisticated our species has become. Beneath and enfolded within the cortex are the various mid-brain limbic structures responsible for modulating our emotional and mammalian experiences. Beneath this is the diencephalon, which modulates and regulates sleep, appetite, and other sensory and homeostatic functions. The very lowest region includes the parenthesis brain. Here the functions of breathing, movement, blood flow, somatic temperature, and other primal functions are regulated.

Neurons, or brain cells, with dendrites and axons interconnect and interpenetrate this vast living web of information and energy and radiate the mystery of consciousness. Neuromelanin is an intimate aspect of this bioluminous information process because of its unique biochemical and bioelectrical properties. Needless to say, this information process not only interconnects the brain within itself and with the body's other organ systems but also with the wider social, cultural, political, and even solar environment. We are photoelectric beings. Each author will explore this phenomenon in a unique way, including its scientific, metapsychological, and bioluminous implications.

There will be six chapters that progressively unfold this subject. The chapters all build upon and draw from each other, seeking in their integration a paradigm shift in this study because of its importance to all of us both medically and neurologically. Future experimental and observational data will deepen this understanding. There is not enough room in the present volume to address the human body's internal organ systems and the role of melanin in health care. Suffice it to say that melanin is present and plays a significant role in the body's internal organ systems, including the heart, liver, gastrointestinal tract, eyes, skin, sexual organs, and the diffuse neuroendocrine system. This edition will focus primarily on neuroscience and the psychological-psychiatric aspects of this vast discipline.

In particular the chapters will stress that melanin and especially brain, or neuromelanin-is present in *all* human beings, internally as well as externally as a matter of complex biology. It will also be noted that melanin and neuromelanin appear, in crucial ways, to guide human embryological development in the womb and the differentiation of the organ systems (Barr, 1983). Because melanin and neuromelanin are ubiquitous in higher life forms, and the "higher" the level of nervous system development among animals, on through the primates, the higher the amount of neuromelanin in the cerebral structures, a unique contribution of this paradoxically "dark" but "light" absorbing substance in evolutionary unfoldment is suggested. This strongly implies that, contrary to past clinical and academic opinion, neuromelanin is not simply a" waste product" of the system but instead represents a crucial, albeit not fully understood, factor in the complex processes of life.

At times, attention will be focused on the fact that the molecular structure of melanin and neuromelanin demonstrates semiconductor properties and indeed may be a biological superconductor under certain conditions. In other words, it has the capacity for luminosity in restricted settings. Given the affinity of human consciousness for light itself, this opens the door to processes that, while rooted in neuroscience, have implications for human consciousness that greatly transcend our current scientific conceptual paradigms and reach into the contemplative traditions and disciplines studied by students of meditation for untold millennia. Melanin is found in the inner workings of our bodies, our organ systems, and our intimate cerebral structures, as well as in the wider solar and cosmic ambience around us. Is there an interface? Is there a connection relevant to human health and psychology?

Professor T. Owens Moore opens with a chapter grounding us in the neural mechanisms underlying these pigmented neurons of the nervous system. Thereafter follows a crucial elaboration of the developmental origins of these neuromelanin neurons and their subsequent distribution throughout the unfolding process of embryogenesis, or development in the womb. The role of neuromelanin in sensory and motor functioning and enhancement is presented in this context. From this perspective the neuro degenerative processes found in both Parkinson's and Alzheimer's diseases are thrown into a new light. The semiconductor and energy transfer properties of neuromelanin are critical in all of these processes. This work represents a continuation of Dr. Moore's earlier work in the *Science of Melanin* (Moore, 1995, 2002).

Professor Ann C. Brown picks up the story by acknowledging that these pigmented organic biopolymers are found not only in the brain but also in the biosphere, atmosphere, lithosphere, and the cosmos itself. Brain melanin is found in crucial areas in the very center of the brain's deep cavities, the mid-brain's circumventricular organs. Dr. Brown points out that these sub cortical pigmented nuclei and their "neuron circuitry" are very responsive to the complex plasticity of the cerebral processes and are also implicated in the actual generation of new nerve cells, a process called neurogenesis. Other brain centers are then explored in this clinical and neuroanatomical context, which, as in the opening chapter by Professor Moore, extends and reframes our understanding of Parkinson's and Alzheimer's diseases. African Blacks and Asians have the lowest incidence of Parkinson's whereas Europeans have the highest levels. What, if any, is the connection here? The influence of diet and nutrition is shown to be crucial in the neurodynamics of neuromelanin.

In a primarily clinical article focused on psychosomatic medicine, Edward Bruce Bynum presents a way for the practicing clinician to intervene in the functions of the emotionally reactive autonomic nervous system to effect anxiety-mediated somatic and behavioral symptoms. The neural and somatic organ system operations are thought to be deeply influenced by these neuromelanin-implicated processes. The role of neuromelanin's semiconductor and bioluminous properties is pivotal. There follows an elaboration of this perspective from both a present day clinical perspective, focusing on the human "emotional brain," or mid brain limbic system, and also its connection with other healing and contemplative traditions from diverse peoples of the earth. Many of these ideas were outlined in Dr. Bynum's research on *Our African Unconscious* (Bynum, 2021). The bioluminous properties of neuromelanin are seen to modulate our connection to the wider biological and solar ecology. The interwoven fabric of space-time itself is implicated in this process.

Finally, in a clinical and theoretical article drawn from the methodology of dynamic psychiatry, Richard D. King explores the historical roots and range of melanin studies from both modem and ancient perspectives. Melanin studies in some form, he contends, have been part of the medical and sacred studies of mankind for millennia, an idea explored in depth in his seminal work *The African Origin of Biological Psychiatry* (King, 1990; 1994). This is what Dr. King terms the Amenta nerve tract of neuromelanin reflected in brain or neural structures and paralleled in the brain stem. His focus is on the first scientific elucidation of this knowledge in the texts and practices of ancient Kemet. Again the macrocosm-microcosm correspondence in the epistemological paradigm of the ancients is seen to be more than relevant to our present day exploration. How inner brain structures reflect outer physical structures in specific ways is the thrust here: "as above, so below; as within, so without," a kind of neurocosmology (Siler, 1990). Dr. King also concludes that evolution is still active and ongoing and, in agreement with Professor Ann Brown, that neurogenesis, the development of new brain cells, is sustained partially under the aegis of a neuromelanin mediated process.

These chapters, Fire Atoms Ignite Inner Vision by T. Owens Moore and The Farther Reaches of Eldership and the Dreamlife of Families' by Edward Bruce Bynum in combination present a view of this ancient but vast and reemerging field that has implications for many areas of research. This knowledge of the intimate psychoenergetic functioning of neuromelanin in evolution and in the neural structures opens the doorway again to cognitive and contemplative disciplines discovered in the early Nile valley civilizations of Kemet that later flowered in the Indus valley of India and eventually throughout the earth. That rediscovery and its bioluminous potential rooted in the neural network and organ systems of the human body is part of the gift this area of science offers to humanity. Each author is mining a small part of this rich mother lode of knowledge and experience for science and perhaps the arts and more. The health of African and African American populations throughout the world is in serious need not only of clinical expertise and resources but also of a bold new conceptual reorganization that takes into account the latest in medical and experimental research as well as the sustained genius of our forbears who studied the deeper mysteries of the heart and the broader sciences of the mind. It is to that living tradition that this volume is dedicated.

REFERENCES

Akbar, N., 1984. "Chains and Images of Psychological Slavery." Jersey City, NJ: New Mind Productions.

Barr, F. E., 1983. "Melanin: The Organizing Molecule," in D.F. Horrobin, ed., *Medical Hypotheses,* vol. 11, 1-140. Edinburgh: Churchill Livingstone,

Bennett, L., 1962. *Before the Mayflower: A History of the Negro in America.* Chicago: Johnson Publishing Co.

Bohm, D., 1980. *Wholeness and the Implicate Order.* London: Routledge & Kegan Paul.

Breasted, J. H., 1984. "The Edwin Smith Papyrus," in *The Classics Medical Library.* Birmingham, U. K.

Bynum, E. B., 1999. *The African Unconscious: Roots of Ancient Mysticism and Modern Psychology.* New York: Columbia University Teachers College Press.

Chandler, W. B., 1999. *Ancient Future: The Teachings and Prophetic Wisdom of the Seven Hermetic Laws of Ancient Egypt.* Baltimore: Black Classic Press.

Copenhaver, B. P. *Hermetica.* London: Cambridge University Press.

Fanon, F., *1967.BlackSkin, White Masks.* New York: Grove Press.

Finch, C. S., 1990. *The African Background to Medical Science.* London: Karnak House.

Ghalioungui, P., 1973. *The House of Life: Magic and Medical Science in Ancient Egypt,* 31. Amsterdam, B. M.: Israel.

Hourning, E., 1986. "The Discovery of the Unconscious in Ancient Egypt," *An Annual of Archetypal Psychology and Jungian Thought,* Spring, 16-28.

King, R. D., 1990. *The African Origin of Biological Psychiatry.* Germantown, TN: Seymour Smith.

-1994. *Melanin: A Key to Freedom.* Hampton, VA: U. B. & U. S. Communications Systems.

Kybalion, 1940. *Hermetic Philosophy.* Chicago: Masonic Publication Society.

Massey, G., 1881. *The Book of Beginnings,* vols. 1, 2. London: Williams and Norgate.

Moore, T. 0., 1995. *The Science of Melanin: Dispelling the Myths* Silver Spring, M.D.: Becham House Publishers.

--2002. *Dark Matters, Dark Secrets.* Redman, GA: Zaman Press.

Pribram, K. H., 1991. *Brain and Perception: Holonomy and Structure in Figural Processing.* Hillsdale, NJ: Erlbaum.

Siler, T., 1990. *Breaking the Mind Barrier: The Artscience of Neurocosmology.* New York: Touchstone Books/Simon & Schuster.

Stampp, K. M., 1956. *The Peculiar Institution: Slavery in the Ante-Bellum South.* New York: Vintage Books.

Tylor, E. B., 1871. *Primitive Culture.* Cited in J. S. Mbiti (1969) *African Religions and Philosophy.* Portsmouth, NH: Heinemann.

Welsing, F. C., 1991. *The Isis Papers: The Keys to the Colors.* Chicago: Third World Press.

"All truth passes through three stages...

First it is ridiculed.

Second it is violently opposed.

Third it is accepted as being self-evident."

-Arthur Schopenhauer, 1788-1860

"By any means necessary.."

-El-Hajj Malik El-Shabazz (Malcolm X), 1925-1965

"From an African-centered perspective, we understand truth to be inseparable from the search for meaning and purpose-the unique concern of human consciousness. As African scholars, it is our responsibility to create systematic theoretical formulations which will reveal the truths that enable us to liberate and utilize the energies of our people. In this view, the self-determinist, the revolutionary, and the scholar are one, having the same objective, involved in the same truth-process. The claim that we make is not to spurious 'objectivity' but to honesty."

-Marimba Ani, *Yurugu: An African-Centered Critique of European Thought and Behavior*

CHAPTER 1

Neuromelanin:
A Highly Sensitized Sensory-Motor Network
T. Owens Moore, Ph.D.

OVERVIEW

This chapter will explore the neural mechanisms underlying pigmented neurons in the nervous system. In general, the dark pigment deep within the brain (i.e., neuromelanin) has a critical role in promoting an optimal level of nervous system functioning. In this chapter, the developmental origin of neuromelanin, its subsequent distribution within the nervous system, and its biosynthesis within the brain will be reviewed. Experimental research on the biophysical properties of melanin has provided significant information on the supportive role of neuromelanin in the nervous system.

Although early research suggested that neuromelanin was a waste product, there is a plethora of recent evidence to dispute this claim. Current research on the biophysical properties of melanin can elucidate the mechanisms by which neuromelanin functions in the nervous system. For example, neuromelanin is an antioxidant that can prevent cellular damage, it can act as a semiconductor by increasing the speed of nerve impulses, and it can function as an electrochemical transducer to transform physical stimuli into neural activity. The electrochemical transducing capability and the overall biophysical properties of neuromelanin appear to be associated with advanced neural processing in mammals and especially primates. In sum, this dark matter in the brain is a highly sensitized sensory-motor network.

INTRODUCTION

Melanin is the general term used to describe pigment in humans. Inhumans, pigments come in various colors that are found inside as well as outside of the human body. This topic has been presented by scholars in numerous fields of study to provide a unique perspective of melanin functioning in humans. For example, chemistry, biology, physics, psychology, and neuroscience are areas of study that have contributed to our understanding of the role and significance of melanin in the nervous system (i.e., neuromelanin).

An extensive review of the science of melanin has been documented by this author (Moore, 1995, 2002) to validate the numerous roles for melanin in human physiology. The literature reviewed in this chapter will be from experimental animal research and research on neuromelanin in the human nervous system. The reader will gain an understanding of the significance of the dark pigment deep within the brain of humans. The origin, location, biosynthesis, and advantages of neuromelanin functioning will be explored.

A MYSTERY UNVEILED

Melanin is a mystery because it is a pigment associated with Blackness. In this race-conscious society, issues related to Blackness can be a lightning rod for controversy. On one hand, the area of melanin research has been neglected because of the politics of scientific research. Since much of the experimental research has been conducted by non-Black scientists, it is this author's opinion that an objective perspective of this area of research has been neglected. On the other hand, if African-centered scientists explore the area of melanin research with provocative intellectual questions, they may be labeled as reverse racists or pseudoscientists (de Montellano, 1993) for raising issues that are difficult for non-Black scientists to comprehend. But the mystery of melanin functioning has been revealed, and we can no longer view it as a waste product in the nervous system (Graham, 1979).

In 1975 Clark, McGee, Nobles, and Weems wrote a classic article and presented a very clear and concise analysis of melanin in the central nervous system (CNS). It is known that our CNS performs a critical information processing role that is essential for optimal neurological and metabolic functioning. According to Clark et al., there is a high, positive correlation between specific levels of sensory activity and states of pigmentation that has been examined by neurophysiological, neurochemical, and neurohumoral data. These authors were led to believe that melanin refines the CNS and, in so doing, produces a highly sensitized sensory-motor network.

According to Clark, et al., the original race was African, and people of African descent exhibit the largest quantity of melanin. The authors make a connection between melanin and CNS functioning by stating that a major portion of the empirical research conducted in African psychology involves a systematic examination of the relationship between melanocytes and the nerve cells of the CNS. As we will discuss in the next section, both are embryologically derived from a single neuroblast in the neural crest of the developing human fetus.

ORIGIN OF NEUROMELANIN

Cells in the skin that produce, synthesize, and secrete melanin are called melanocytes. Melanocytes, which are found in the basal cell layer and between cells of the epidermis, have dynamic functions because they can change as a result of direct physical stimulation or from the normal aging process. In contrast, pigmented neurons in the nervous system are not commonly named melanocytes because they are less dynamic and more static. Neuromelanin is found in the cytoplasm of pigmented neurons (Bazelon, Fenichel, and Randall, 1967), and the presence of neuromelanin is less subject to change.

The amount of neuromelanin in the nervous system is determined by genetics. It is quite clear that the amount of external melanin varies between ethnic groups; however, it is considerably more difficult to measure ethnic variations in the stable internal melanin. Although the terms static or stable are used here, it is not implied that neuromelanin levels can never change in an individual. For example, deterioration of neuromelanin-containing brain cells can be caused by drugs, chemicals, or the normal aging process.

The primary role for melanin in the skin is to protect against ultraviolet radiation. Assuming that ultraviolet radiation does not penetrate the brain, neuromelanin in the nervous system and melanin in the skin have slightly different functional roles. In contrast, an analysis of the origins of internal and external melanin can demonstrate the commonality between the two forms of melanin.

Developmental Embryology

During fertilization, a zygote forms and divides into three distinct germ layers: the endoderm; the mesoderm; and the ectoderm. To understand the important role of melanin during the early stages of embryonic development, we must focus on the derivatives of the outer embryonic layer-the ectoderm. The ectoderm is composed of three regions: the prospective neural tube; the prospective neural crest; and the prospective epidermis. It is within these three regions that melanin plays its first key role in maintaining life. In the midst of this embryological darkness, an explosion (Big Bang) will occur that will transform the tiny mass of cells into a complex human being.

Each region of the ectoderm is further differentiated into specialized body parts. The *neural tube* is differentiated into the following parts: a) the brain; b) the posterior pituitary gland; c) the optic vesicles; d) the spinal cord; and e) the motor nerves that originate in the ventral portion of the neural tube and innervate muscles.

The *neural crest* derivatives consist of cells that migrate to distant parts of the body. These migrating cells form sensory nerves and ganglia, which receive impulses from the following sites: a) sense organs; b) autonomic ganglia; c) the adrenal medulla; d) all of the pigmented retinal cells, which are derived from the neural tube; e) the cartilages in the voice box and head; and f) some of the ectodermal muscles.

The *epidermal layer* can be divided into cells derived from epidermal thickenings and those derived from the rest of the epidermis. The thick epidermal derivatives include some of the cranial nerves, the lens of the eye, the olfactory structures, the inner ear, and the taste buds. The remainder of the epidermis forms the following structures: a) the outer layer of the skin; b) the hair and nails; c) the linings of the mouth and anus; and d) the anterior pituitary gland (Oppenheimer and Lefevre, 1984).

Melanin is in numerous locations in the body, and the importance of neuromelanin in brain tissue will be discussed in detail by Dr. Brown in Chapter 2. For this chapter, the reader need only be aware of neuromelanin's role during embryogenesis.

Neuromelanin and Embryogenesis

Melanin and neuromelanin are part of the sensory-motor network from the earliest stages of embryogenesis. As mentioned in the previous section, the brain and the spinal cord are formed from the neural tube, the sensory-motor network extends from the neural crest, and melanocytes in the skin come from the epidermal layer. Each site is dependent on the presence of melanin for proper physiological functioning.

Since our sensory apparatus is so vital to learning, it begins to develop *in utero* within a couple of months after conception (Hannaford, 1995). Nerves appear three weeks after the egg is fertilized and immediately begin to link up with other nerves. Before birth, we learn about gravity through our vestibular system, and our body becomes a fine-tuned sensory receptor for collecting information. During embryogenesis, our sensory-motor system shapes our experience, and we are shaped by the events that occur *in utero.*

The human embryo develops out of a womb of **darkness**. There is no sunlight available to stimulate the production of melanin inside the fetus or on the external surface of the fetus. All sources of melanin migrate to their destination sites due to chemical factors such as neurotrophins (Reichardt and Farinas, 1999). Neurotrophins expressed in targets can promote survival of neurons whose cell bodies are distant ganglia.

Experimental research has shown that extirpation and explanation of the neural crest reveal that it is the sole source of all pigment cells in the body, except those that differentiate in the retina and are therefore derived from the optic cup (LeDouarin and Kalcheim, 1999). If portions of neural crest cells are removed, body parts would be missing, and pigmentation would be absent. On the other hand, grafting tissue from the neural crest onto the embryo can increase pigmented cells.

In sum, the aggregation of heavily melanized cells formsthe grossly visible black pigmentation in the developing embryo. Melanin and neurotrophic factors move into the folds of the neural groove and appear to bring the folds of the neural tube together to form the brain and the spinal cord. The formation of the neural crest helps to fuse the neural folds, and the nervous system begins to take form and shape.

Sensory Enhancement

The nervous system is divided into the CNS and the peripheral nervous system (PNS). The brain and spinal cord are the main components of the CNS, while the PNS is further divided into the autonomic and somatic nervous systems. The processing of sensory information occurs in every aspect of the nervous system, and the presence of melanin can influence both the development of the nervous system as well as how brain cells integrate sensory information with other physiological systems. For example, the migration of cells to make the melanocytes in the skin, sheds light on the skin as an extension of the nervous system.

Melanocytes are multidendritic cells derived from the neural crest that look very similar to the structure of nerve cells in the brain. Fig. 1 illustrates a comparison of the two cell types. Neurons in the brain and melanocytes in the skin have the same embryological origin, and previous work (Moore, 2002) has presented experimental evidence to support the hypothetical role of the skin as a large brain. In addition, glia are the other major components of the nervous system, and the morphology of glial cells is very similar to that of melanocytes.

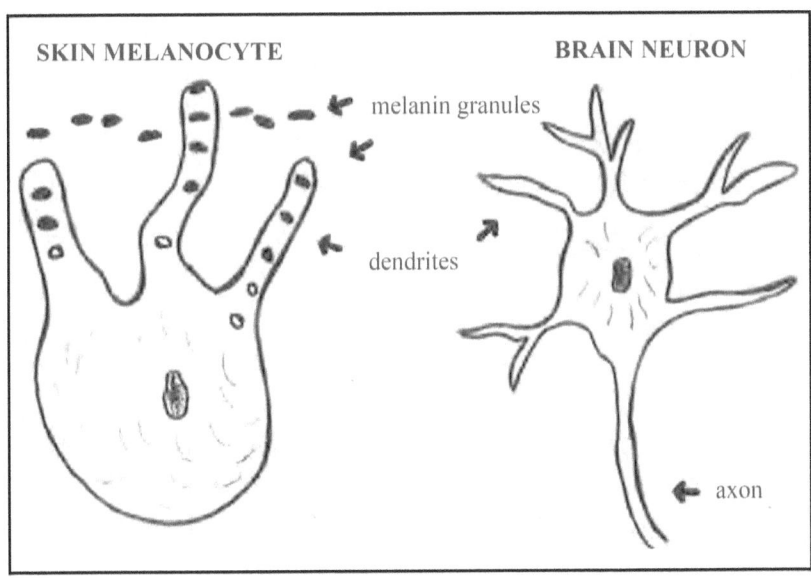

FIG 1 -A comparison between the morphology of a skin cell (i.e., melanocyte) and a brain cell (i.e., neuron). Both cell types express dendritic processes that are activated by external stimuli. Both cells secrete chemicals (e.g., melanin by the melanocyte and various types of neurotransmitters by the neuron).

Moreover, melanin migrates to the peripheral nervous system, where the nervous system is in direct contact with the external world via the autonomic nervous system (nerve ganglia connecting all internal organs) and the somatic nervous system (musculature). The endocrine system, which secretes various hormones, is also dependent upon melanin for its structure and function. Furthermore, the nerves for olfaction, vision, and hearing are all formed from the presence of melanin, and neurons in these areas contain melanin granules in their cytoplasm. Melanin in all of these regions of the human body can help to provide extrasensory perceptive abilities. Life is the experience of perceiving what nature has to offer, and experimental research on the biophysical properties of melanin suggests that the life experience can be enhanced by the presence of melanin.

To clarify this latter point, conceptualize the following three factual scenarios: 1) the absence of pigment in the inner ear would produce deafness; 2) the absence of pigment in the retina would make us virtually blind; and 3) without pigment in specific mid brain structures, our psychomotor abilities would be severely impaired. Next, we will review the location of neuromelanin in specific brain structures.

LOCATION OF NEUROMELANIN

The distribution of melanin in the human brain has been mapped by numerous investigators (Olszewski and Baxter, 1954; Olszewski, 1964; Bazelon, et al., 1967; Fenichel and Bazelon, 1968; Graham, 1979; Van Woert, Prasad, and Borg, 1967; Sapper and Petito, 1982; Bogerts, 1981), the cells being primarily located in the brainstem and the mid-brain.

Bogerts (1981) studied the brains of four adults (ages 47, 53, 54, and 56) and demonstrated a striking similarity between the location of melanin and the catecholamine cell bodies described in various animal and human fetuses. From the brain areas that were studied, cell counts from the center of each area showed that the mean density of melanin-containing cell bodies varied considerably among the different areas.

Another significant finding from this study was that neuromelanin was not found in the dopamine mesolimbic pathway that is responsible for the reward or pleasure circuit extending from the ventral tegmentum to the nucleus accumbens. This lack of neuromelanin in the reward pathway was surprising because of the concept that both neuromelanin and catecholamines are commonly derived. Others have reported similar findings with a rat that was genetically sensitive to dopamine (Rot, et al., 1995). There were changes in the nigrostriatal dopamine system but not the mesolimbic dopamine system.

These are intriguing findings because the great majority of catecholamine melanin-containing cells are dopaminergic (Breathnach,1988). It is known that melanin is deposited in neurons that are most active in catecholamine synthesis, and the accumulation begins early in life and increases as the animal ages. Hence, neuromelanin may be a marker of active catecholamine metabolism (Fenichel and Bazelon, 1968).

The distribution of melanin-pigmented neurons in the human brain closely corresponds to that of catecholamine cell groups in the brains of other species (Breathnach, 1988). Furthermore, there are species differences in neuromelanin containing cells, with guinea pigs and rabbits having none; rat, cat, and dog, some; higher primates more than lower primates; and man with the most of all (Breathnach, 1988). There are also differences according to age distribution in man (Mann and Yates, 1983; Marsden, 1983), and these differences suggest that neuromelanin is not toxic *per se* or just an end product of the biosynthesis of catecholamines.

Early reports suggested that pigmentation was not found in substantia nigra of rabbit, rat, mouse, guinea pig, or most marsupials (Marsden, 1961). In addition, intensity is age-dependent since pigmentation was lacking in the substantia nigra of young animals though melanin granules were reported in the human locus coeruleus as early as the fifth month of gestation (Foley and Baxter, 1958).

Other than age as a likely factor that can reveal differences in the level of neuromelanin, it is very significant to mention how ethnic differences are overlooked when evaluating the presence of neuromelanin. If the claim is made that brain melanin is programmed to function at different capacities depending upon a person's overall genetic capacity to produce melanin, then it is likely that brain melanin can vary between ethnic groups (Moore, 1995).

The distribution of melanin-pigmented neurons in the human brain was plotted using the brains of three adults, 68 to 80 years of age (Saper and Petito, 1982). These neurons were found primarily in areas corresponding to the A1-A14 catecholamine cell groups that have been reported in other species (Dahlstrom and Fuxe, 1964; Garver and Sladek, 1975). The cells from A1 in the reticular formation extend through the brainstem and mid-brain to the A14 cell groups in the hypothalamus (Saper, et al., 1982).

In another display of the cytoarchitecture of the human brainstem, Olszweski and Baxter (1954) mapped 12 neuromelanin cell groups, which can be found in Table 1. The substantia nigra and the locus coeruleus are the two major neuromelanin-containing cell groups. The substantia nigra produces dopamine and the locus coeruleus produces norepinephrine. To provide a visual representation of the location of neuromelanin, Fig. 2 is an illustration of the defined area of the substantia nigra and the locus coeruleus in the brain of a rat immunocytochemically stained for tyrosine hydroxylase. Tyrosine hydroxylase is an enzyme involved in the biosynthesis of catecholamines.

There is another prominent structure called the red nucleus in the mid-brain area that is visually distinct as a collection of pigmented neurons. In brief, the red nuclei on the left and right sides of the mid-brain send axons across the midline of the brainstem into the spinal cord. The structure running from the red nucleus to the spinal cord is called the rubrospinal tract, and experimental evidence has demonstrated that this system plays a significant role during voluntary movements of the arm, hand, and fingers (Squire, Bloom, McConnell, Roberts, Spitzer, and Zigmond, 2003).

In the next section, we will briefly explore whether or not there is a role for catecholamines in the biosynthesis of neuromelanin.

TABLE 1- 12 Neuromelanin Cell Groups in the Human Brain Stem Mapped by Olszewski and Baxter (1954) using the Nissl staining technique. Cytoarchitecture of the Human Brain Stem by S. Karger: New York.

1. **SUBSTANTIA NIGRA (SN)** - Dopamine-containing cells
2. Nucleus parabrachialis
3. Nucleus paranigralis
4. **LOCUS COERULEUS (LC)** - Norepinephrine containing cells
5. Nuclei intracapsularis
6. Subceruleus
7. Nervi trigemini mesencephalius
8. Pontis centralis oralis
9. Tegmenti pedunculopontinus
10. Parabrachialis medialis
11. Dorsomotor
12. Retroambigualis

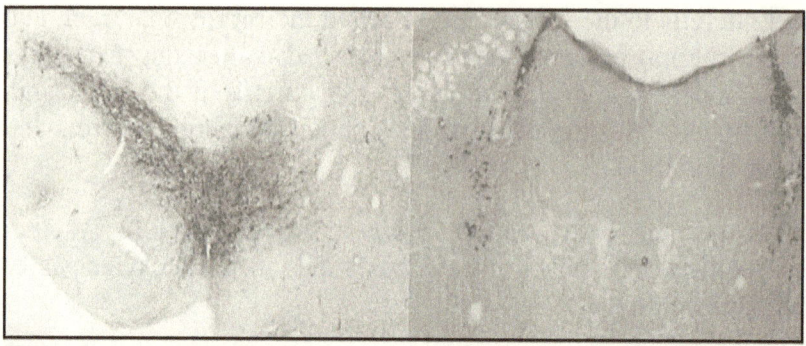

FIG 2 - *An immunocytochemical technique was used to stain for tyrosine hydroxylase in the substantia nigra (left view) and the locus coeruleus (right view) of a rat brain. The collective group of neurons in each nucleus reveals a distinct section of the brain that would contain neuromelanin.*

BIOSYNTHESIS OF NEUROMELANIN

The biosynthesis of neuromelanin is a fascinating phenomenon because it does not appear to form in the same manner as skin melanin. Skin melanin strictly relies on enzymes (e.g., tyrosinase), whereas little to no evidence supports enzymic activity in pigmented brain cells. For example, Rodgers and Curzon (1975) used a quantitative radiometric assay to compare various substances as melanin precursors in the rat brain. Their goal was to study the ability of different brain regions to form melanin and to evaluate various hypotheses of brain melanin formation. The results of their findings demonstrated that catecholamines, L-DOPA, and serotonin were precursors for brain melanin formation. The assay was used to evaluate various hypotheses of brain melanin formation. However, no evidence for enzymic activity was found, and it was concluded that brain melanin formation might be a largely nonenzymic process. Tyrosinase, peroxidase, and monoamine oxidase activity were all investigated, and there was minimal to no activity for these enzymes in the brain regions studied. Although enzyme activity appeared to be absent, melanin formation was detected in all brain regions studied and was highest in the substantia nigra and the striatum.

In studies examining the etiology of Parkinson's disease (Goldman and Tanner, 1998), the commonly held belief is that neuromelanin is formed by dopamine autoxidation. This concept of autoxidation was proposed by Doyle Graham in 1979. Even with a lack of external melanin, the autoxidation mechanism is a likely explanation for the presence of neuromelanin in a person who lacks substantial amounts of external melanin. However, the inclination is to accept neuromelanin as a waste product of the autoxidation mechanism.

In contrast, Barr (1983) stresses the point that neuromelanin is capable of self-synthesis. The self-synthesis hypothesis can buttress the view that neuromelanin is genetically programmed to function at a different capacity, depending upon a person's overall genetic capacity to produce melanin in the body (Moore, 1995).

GENETICS AND NEURODEGENRATIVE DISORDERS

The human genome project has provided valuable information related to diseases that manifest in humans. From current genetics research, there is strong evidence linking genetic abnormalities to the development of neurodegenerative disorders such as Parkinson's disease (PD) and Alzheimer's disease (AD). For example, mutations in the alpha-synuclein gene are implicated in PD (Singleton, et al., 2003). Alpha-synuclein was identified as a major component of Lewy bodies, the pathological hallmark of PD and glial cytoplasmic inclusions (Tu, et al., 1998). A similar disease process may resemble the etiology of AD in Down's syndrome, where expression of the amyloid precursor protein (APP) gene due to chromosome 21 trisomy is the event (Monsonego and Weiner, 2003).

The pathology associated with both PD and AD may be associated with the accumulation of toxic substances over the years. In AD, amyloid beta-peptide is a cleavage product of neuronal APP. Genetic factors such as APP or presenilin mutations or carrying of the ApoE4 allele results in earlier accumulation of amyloid beta-peptide and early onset of clinical symptoms (Monsonego, et al., 2003). In both PD and AD, it is unclear what crucial factors in aging trigger the accumulation of toxic substances in the brain.

In sum, a person can have a genetic predisposition to develop neurodegenerative diseases if genetic mechanisms are unavailable to ward off the accumulation of neurotoxic substances. If a person has a proclivity to develop a neurodegenerative disease, then several factors (e.g., age, diet, toxic exposure) may help to trigger the destruction of neurons. Therefore, it is hypothesized that neuromelanin is genetically programmed to enhance neural functioning by protecting the brain from assaults by toxins that accumulate over years of slow poisoning.

NEUROMELANIN AND ETHNICITY

For obvious reasons, it is not possible to dissect the brain of a living human in order to explore how neuromelanin functions, but the biophysical properties of synthetic melanin in the laboratory can shed light on how neuromelanin functions in the brain. In addition, evidence can be gathered from animal experiments, pharmacogenetic effects of drugs in patients, and post-mortem analysis of brain tissue. Racial classifications have historically used skin pigmentation as the sole factor in making distinctions between groups of people. The different levels of skin pigmentation are due to enzymes (e.g., tyrosinase) that determine the amount of melanin that is produced on the external surface of the body. No controversies arise in discussing ethnic differences in skin melanin, but the subject becomes problematic when discussing ethnic variations in brain melanin.

Nearly two decades ago, Lawson (1986) raised the issue of racial and ethnic factors in psychiatric research. There was no direct correlation with melanin or neuromelanin, however. We are raising the issue in this book since, as previously stated (Moore, 1995), brain melanin is genetically programmed to function at different capacities depending upon a person's overall genetic capacity to produce melanin.

While the internal molecular arrangements and diffuse interconnections of one's neuromelanin-catecholamine network may vary uniquely with each individual, such variations in one's neuromelanin do not correlate (in any obvious way) with one's skin color, whether white, red, yellow, black, brown, or albino (Barr, 1983). A special character of brain melanin is suggested by the normal pigmentation of the substantia nigra and locus coeruleus of albinos who lack melanin pigments elsewhere (Foley and Baxter, 1958). Another intriguing characteristic is the absence of reports of melanomas of these pigmented brain regions even though melanomas of other melanized tissues are well known (Curzon, 1975). To make sense of these anomalous observations, Barr (1983) reminds us that most studies on neuromelanin have been with synthetic melanin *in vitro*. He further states that definitive studies should be done *in vivo* (Barr, 1983).

Definitive studies would help to do away with the concept that neuromelanin is only a waste product. It is the opinion of this author that the waste product terminology should be discarded for two specific reasons: 1) it implies that neuromelanin has no significant value; and 2) it implies that neuromelanin should be the end-product formed wherever catecholamines are in high concentration. The first point will be addressed in the next section to emphasize the multiple advantages of neuromelanin functioning. The second point pertains to enzyme activity, which other reports have addressed (Curzon, 1975). Other factors must play a determining role for the presence of neuromelanin, and the field of genetics can provide clues. Moreover, the biochemistry of melanin/ neuromelanin may be the critical factor linked to any genetic variations in the manifestation of psychiatric disorders.

For example, Lawson (1986) has brought to our attention that racial and ethnic differences exist in the symptom presentations of psychiatric disorders. In addition, clinical issues have been raised (Lawson, 1996a) to suggest that African Americans have poorer outcomes than Caucasians in general health and mental health systems, possibly due to lesser access to services, particularly psychopharmacology in mental health systems. There was only a suggestion that these problems may be exacerbated by ethnic differences in pharmacokinetics (Lawson, 1996a, 1996b).

A research team in the Netherlands has uncovered some revealing findings that connect genetics and neural sensitivity in rats (Rot, et al., 1995). Although it is an animal experiment, the data can indirectly support the claim in this chapter that neuromelanin is genetically programmed to function differently depending upon the amount of melanin produced by the individual.

In this experiment, pharmacogenetically selected Wistar rat lines were used to investigate the implication of either high or low responsiveness of the dopamine system to the activity of the hypothalamic-pituitary-adrenal (HPA) axis. Apomorphine, a dopamine agonist, was the pharmacological agent used to measure differences between apomorphine-susceptible rats (apo-sus) and apomorphine-unsusceptible rats (apo-unsus).

The findings revealed increased binding of the dopamine antagonist iodosulpiride to D1/D3 receptors and increased D1 and D2 receptors and mRNA expression in the striatum of apo-sus rats. Moreover, apo-sus rats expressed higher tyrosine hydroxylase mRNA levels in the substantia nigra pars compacta. Collectively, these markers suggested a higher biosynthetic capacity of dopamine in the nigrostriatal pathway and a higher responsiveness of the striatal dopamine receptors. This potentially enhanced reactivity of the nigrostriatal dopamine system coincides with the much higher sensitivity of these rats to apomorphine-induced stereotypic behavior.

In addition, the authors demonstrated a genetic link between dopamine susceptibility and stress-induced HPA activation. For example, their findings suggested that increased susceptibility of the dopamine system, and increased tyrosine hydroxylase mRNA and D2/D3 receptor capacity, coincides with increased central HPA drive and corticosteroid feedback resistance. Since it was demonstrated that there are differences in neuroendocrine response patterns against a dopamine dependent genetic background, it is hypothesized that individual differences in dopamine are linked to the magnitude and duration of HPA activation. In other words, a more sensitive dopamine system can help an individual effectively deal with stress. The dopamine system is activated during stress to release HPA hormones, but overactivation may lead to negative physiological consequences. The high incidence of stroke, hypertension, and other cardiovascular imbalances could be the result of overly sensitive dopamine systems that impact the neuromelanin catecholamine network.

Ethnobiological differences in response to drugs have been observed with both psychoactive, or mind-altering drugs (Lin, Poland, and Chien, 1990; Strickland, Lin, Fu, Anderson, and Zheng, 1995; Wood and Zhou, 1991), and nonpsychoactive drugs (Flaherty and Meagher, 1980). Strickland and Gray (2000) explored the significance of ethnobiological variation in drug responsivity to drugs that are known to treat mood disorders such as depression, anxiety, and schizophrenia.

Many of the drugs that have been developed to treat mental disorders affect neurotransmitter systems that come from melanated centers in the brain. For example, norepinephrine, serotonin, and dopamine are neurotransmitters involved in mood disorders, and each neurochemical is produced from specific brain nuclei that are visibly melanated. Interestingly, it has been reported that African Americans have a higher risk of developing tardive dyskinesia than Caucasians, even when differences in neuroleptic drug use are accounted for (Eastham, Lacro, and Jeste, 1996). Asians appeared to have a lower or equal risk of developing tardive dyskinesia compared with Caucasians.

Even though scientists claim that the amount of neuromelanin is independent of skin melanin, there is reason to believe that there could be variations in the sensitivity of neuromelanin functioning. Everyone has skin covering their bodies, but some people are more sensitive to touch. Everyone has eyes to see, but some people are more sensitive to certain types of light. Everyone has ears, but some people are more sensitive to specific kinds or frequencies of sound. Likewise, one can have a genetic predisposition for <u>a more sensitive neuromelanin catecholamine network.</u>

Parkinson's Disease

Parkinson's Disease (PD) is a neurodegenerative disease that mostly affects the elderly. PD is pathologically characterized by destruction of dopaminergic cells in the mid-brain region called the substantia nigra. The substantia nigra is a highly melanated subcortical structure. As part of the basal ganglia (motor system), it has an integral role in regulating and producing slowly coordinated and deliberate movements. Besides destruction of substantia nigra neurons, in PD there is also degeneration of cells in the ventral tegmentum and the locus coeruleus (Kaplan and Sadock, 1987). The combined effect of this neurodegeneration leads to tremor, muscle rigidity, bradykinesia (slow movements}, stooped posture, and a shuffling gait. By observing the psychomotor impairments of neurodegenerative disorders such as PD, one can infer how the brain controls normal behavior.

A thorough investigation of the behavioral impairments and motor deficits associated with PD can indirectly reveal how the substantia nigra functions. It can be speculated that a highly melanated substantia nigra can produce advanced motor skills throughout the lifetime of the organism. As the individual ages, the presence of neuromelanin increases, but there may be a loss of catecholamine neurons in the melanated brain region. Age-related factors have been the primary contributors to the development of PD; however, there are numerous cases of PD-like symptoms in young people who have experimented with drugs. In the 1970s there was a connection between PD and a synthetic drug called MPTP, and 30 years later a connection was established with another designer drug called ecstasy. Therefore, age is not the only etiology for PD. Drug effects on the pigmented neurons in the brain suggest that the biophysical properties of neuromelanin can contribute to the destruction of these specialized cells that are primarily involved in psychomotor tasks.

The effects of designer drugs in young people have led some to believe that there are environmental risk factors associated with PD. Marder and colleagues (1998) used univariate and multivariate unconditional logistic regression models in 89 nondemented patients with PD and 188 control subjects in a multiethnic urban community. Rural living, area farming, and drinking well water were associated with PD only in African Americans. In Hispanics, area farming was protective, whereas drinking unfiltered water was a risk factor for PD. The authors concluded that ethnic and cultural origin might add to the epidemiological study of PD.

There has been a suggestion that people of African descent are less likely to develop PD when compared to other ethnic groups. For example, the large-scale pattern of underlying-cause PD mortality among Whites has persisted for three decades (Lanska, 1997). Lanska used the National Center for Health Statistics and the Bureau of the Census to map age-adjusted, race- and race gender-specific PD mortality rates in the U. S. for 1998. Reported rates among Blacks were significantly lower than among Whites. Among Whites, high underlying-cause rates predominated in the North and low rates predominated in the South.

In 1993, Gilbert analyzed the low risk of aging Africans as opposed to the high risk of Caucasians to certain major disorders. The disorders included PD, myocardial infarction, osteoporosis and fractures, some rheumatic diseases, and cancer. In this European-based study, the relative risk was determined by a common physiological mechanism involving the autonomic nervous system and calcium metabolism. In addition, increased vagal tone, enhanced dopaminergic activity, an efficient dopamine/vitamin D parathormone system, and a neuroendocrine-metabolic context could determine the response to specific stimuli.

As speculated from the previously mentioned genetic sensitivity study with rats (Rot, et al., 1995), Gilbert suggested that maintained dopaminergic activity, as proposed for Africans, coupled with low risk to certain disorders confirms the experimentally demonstrated paramount importance of this neurotransmitter in retarding aging processes in animals. The neuroendocrine profiles as defined for Africans is consistent with a potentially extended period of physical and mental competence and a conceivable shorter duration of involuntary decline (Gilbert, 1993).

In another epidemiological study, Kurtzke and Goldberg (1988) reported that age-adjusted death rates for PD in the U.S. from 1959 to 1961 demonstrated significantly lower rates for Blacks than for Whites, with rates for Asian Americans the same as for Whites. In another study (de la Monte, Hutchins, and Moore, 1989), it was determined that the dementia due to PD was more frequent among Whites; the frequencies of multi-infarct dementia and dementia due to chronic ethanol abuse were higher among Blacks; and the frequency of Alzheimer's disease was 2.6 times higher among Whites. The study explored racial differences in the etiology of Alzheimer's disease, and a strong argument was made in favor of genetic transmission of sporadic Alzheimer's disease.

Some studies report that Blacks have a lower rate of PD than Caucasians, whereas other studies have not found such a difference. For instance, a review conducted by Richards and Chaudhuri (1996) reported that the confounding effects might be due to a low case ascertainment and high selective mortality. Even though people of African origin may be more protected from the effects of PD, there is data to suggest that people of African descent are vulnerable to vascular PD, which is associated with high mortality. Moreover, lower life expectancy and failure of old people to attend hospitals in South Africa may be factors in the apparent low prevalence of PD among Blacks (Cosnett and Bill, 1988).

A study performed in India (Muthane, Yasha, and Shankar, 1998) further supports the ethnic variation in neuromelanin that has been reiterated in this chapter. The objectives of this study were to count the number of melanized neurons in the substantia nigra pars compacta in normal human brains from India, study the change in neuronal count with advancing age, and compare the neuronal counts from this Indian population with counts reported in normal brains from the United Kingdom (UK). In the brains from India, there was no loss of melanized nigral neurons with advancing age. The absolute number of these melanized neurons was about 40 percent lower than in the brains from the UK. Despite a low number of melanized nigral neurons in the brains from India, individuals function normally and have dopamine levels comparable with their Western counterparts. Therefore, it is not solely the absolute number of melanized nigral neurons that attributes to the development of PD. There is no significant loss of pigmented nigral neurons with age, suggesting that the loss seen in PD is exclusively due to the disease process itself. The authors concluded that Indians have a lower prevalence of PD despite having a low count of melanized nigral neurons, suggesting that better mechanisms to prevent the loss of nigral neurons may be present in Indians.

According to Chaudhuri, Hu, and Brooks (2000), ongoing studies in India suggest that the pattern of PD there tends to differ from Afro-Caribbean subjects in the UK. These authors are attempting to unravel the mechanism of increased frequency of atypical PD in these ethnic groups, and they include genetic studies addressing polymorphisms of enzymes metabolizing levodopa, dietary neurotoxin screening, and functional imaging studies of the striatum using positron emission tomography. As we have discussed earlier, cardiovascular challenges such as diabetes and hypertension are being considered as significant factors affecting people of African descent.

In sum, there is substantial evidence to support the claim that neuromelanin may function differently depending upon biochemistry and a person's overall genetic capacity to produce melanin. Animal experiments exploring genetics and neural sensitivity, ethnobiological variations in response to drug effects, and the higher incidence rate of PD in non-Black people suggest that neuromelanin has a functional role in the nervous system.

ADVANTAGES OF NEUROMELANIN

We commonly think of melanin as a photoprotective pigment in the skin that can block the damaging effects of the sun's ultraviolet radiation (UVR). UVR is an external stimulus, and it does not reach the inner depths of the brain. Therefore, it is difficult to comprehend why melanin is found deep in the brain. Most of the speculation on the significance of neuromelanin formation has resulted from studies of neurodegeneration of pigmented brain neurons or from studies of synthetic melanin in the laboratory. These clueless adventures have slowly faded as we have gained more knowledge concerning the biophysical properties of melanin.

Toxic Neutralizer

Melanin is effective as a means of radiation-less conversion of the energy of harmfully excited molecules into innocuous vibrational energy (McGinness and Proctor, 1973). This conversion may deactivate metabolically excited molecules that can become potentially damaging to cells. Since there are no mechanical devices involved, this conversion is called radiation-less. The important point is that melanin helps to ensure that the spread of further cellular damage is neutralized.

Melanin is also known to protect against dangerous free radicals. Free radicals are highly reactive chemical species that have an odd number of electrons, hence, one unpaired electron. It has been proposed (Commoner, Townsend, and Pake, 1954) that melanin acts as a deposit site or sink for unpaired electrons, thus removing reactive free radicals. Peroxides are examples of chemical substances that can lose an electron and change into dangerous and cytotoxic substances.

Besides cytotoxic molecules and free radicals, melanin has a redox capacity to prevent "rust" in the brain. For example, if you leave iron in water, it will oxidize and rust. In the body, melanin can act as an electron-transfer agent to protect cells and tissue against reducing or oxidizing conditions (van Woert, 1968; Gan, 1976, 1977). By assisting in the transfer of electrons, melanin ensures the safe conversion of potentially volatile chemical reactions.

Nerve Conduction Facilitator

Ions are charged particles that we consume in our diets in the form of electrolytes. When dissolved in water, these electrolytes can conduct an electrical current and set up a weak biological current in the nervous system, which ranges from 40 to 120 millivolts and can be harnessed by melanin (McGinness, Cory, and Proctor, 1974). Melanin can act as a threshold switch to change the voltage for neuronal firing. In other words, there is more activity elicited from melanated brain regions than from nonmelanated brain regions.

It was first suggested by McGinness in 1972 that melanin may act as an amorphous semiconductor since there is a rise in the conductivity of melanin under an applied voltage. McGinness suggested that this rise might be a result of the increased kinetic energy of the electrons, leading to higher mobility and promotion to excited states.

It is suggested that the bioelectronic properties associated with melanin can help to facilitate nerve conduction in the following three ways: 1) it can speed the pace of the nerve impulse; 2) it can concentrate ions for high-voltage generating activity; and 3) it can provide an electrochemical surge.

In the context of mental health, melanated cells can cause a greater release of neurochemicals from nerve cells. Neural transmission requires the stimulation of certain electrolytes across the cell membrane. As a result of this change in the cell membrane, neurochemicals are released from the cell to transmit a nerve impulse. As we previously discussed, there are several brain regions (e.g., substantia nigra, locus coeruleus, raphe nucleus) that have both high melanin content and neurochemical activity (e.g., dopamine, norepinephrine, serotonin). Therefore, neuromelanin can increase the voltage, cause an electrochemical surge, and positively influence the release of neurochemicals.

Energy Transformer

The consistent appearance of melanin in living organisms at locations where energy conversion or charge transfer occurs (e.g., skin, retina, inner ear) is of particular interest. Melanin is strategically located in the body to absorb and to convert various forms of electromagnetic energy into energy states that can be used by the nervous system. Essentially, it can function as an electrochemical transducer.

The fact that melanin is black or dark in color could help explain how it functions as a converter of energy. Since dark skin or any black substance absorbs heat, light is not reradiated but instead converted to rotational and vibrational degrees of freedom (McGinness and Proctor, 1973). Contrary to blackness, whiteness reflects light. Thus, pigmented cells are more capable of converting energy versus nonpigmented cells. Although there appears to be comparable distributions of neuromelanin in the brain when ethnic groups are compared, darker people have a more modified external, or skin, pigmentary system that has a greater capacity for charge transfer.

At the level of the brain cells, however, there are significant neurophysiological functions that are constantly improving the excitation and conductivity of the nervous system. Stimulation of the nervous system requires an action potential, and any change in the electronic nature of the neuromelanin could generate vibrational energy capable of affecting nerve impulses. Action potentials are generated when ions or electrolytes flow in and out of the cell membrane. When physical stimuli from the outside world are converted into neural impulses, neuromelanin could act as a semiconductor to increase the firing rate of action potentials.

BIOLOGICAL ACTIVATION OF NEUROMELANIN

Neuromelanin is a potent antioxidant. To biologically activate neuromelanin and promote the functioning of the neuromelanin-catecholamine network, it is important to ingest the proper nutritional items. Current research has documented the protective role of pigments in certain fruits that can function as effective antioxidants. For example, blackberries and blueberries are at the top of the list of fruits that function as effective antioxidants. Interestingly, the darker the berry, the greater the antioxidant effect. Since neuromelanin has a protective role as an effective antioxidant, one's health can be greatly enhanced by consuming substances that have a similar role in health promotion.

It is essential to consume as many natural foods as possible in order to decrease the chances of developing neurodegenerative diseases. The accumulation of toxic chemicals and artificial substances over the years can be absorbed by the pigmented neurons in the brain and lead to cellular damage. The antioxidant properties found in many fruits can slow the aging process and prevent cellular damage. In addition, studies on the effects of blueberries have demonstrated that these dark berries can improve cognitive skills and motor performance in rats (Joseph, Nadeau, and Underwood, 2002).

In sum, there are many sun-enriched products (e.g., fruits and vegetables) that contain natural chemical substances necessary for the nervous system to function at an optimal level. Vitamins, minerals, amino acids, and alkaloids are a few of the substances that can biologically activate neuromelanin and promote positive physical and mental health.

CONCLUSION

Neuromelanin can be considered the core of consciousness. It is the connection between the blackness of interstellar space (dark matter) and the pigment deep within the recesses of our nervous system (neuromelanin) that allows us to contemplate a new way of thinking about consciousness. As we have pointed out in this chapter, the presence of neuromelanin increases from lower to higher animals and it is highest in man. In terms of conscious awareness, man is usually considered to head the list among species.

The unique display of flamboyant expressiveness in African/Black culture is primarily due to a highly energized sensory-motor network. This expressiveness can be displayed in rhythmically oriented tasks such as dancing, rapping to a beat, moving the body to a percussive rhythm, and nearly all tasks involving psychomotor skills. Beyond skill development, Dr. Bynum in Chapter 3 and Dr. King in Chapter 4 will further discuss the role of neuromelanin in mental processes, or what we know as consciousness. They have made neuromelanin a central theme in their explorations of mind and human consciousness (King, 1990; Bynum, 1999).

From a biological perspective, the presence of neuromelanin can be a double-edged sword. It can protect the brain as well as lead to damaging alterations in brain cells (Faucheux, Martin, Beaumont, Hauw, Agid and Hirsch, 2003). On the one hand, a properly functioning neuromelanin-catecholamine network can greatly enhance a person's psychomotor abilities, but any impairment of these pigmented neurons can cause neurodegenerative diseases such as Parkinson's. To conclude, neuromelanin is not a waste product. Neuromelanin is critical to maintaining a healthy state of mind. To promote positive mental health, the consumption of natural products to stimulate the antioxidant, semiconductive, and electrochemical transducing properties of neuromelanin is encouraged.

REFERENCES

Barr, F. E., 1983. "Melanin: The Organizing Molecule, in D. F.Horrobin, ed., *Medical Hypothesis,* vol. 11, 1-140. Edinburgh: Churchill Livingstone.

Bazelon, M., Fenichel, G. M., and Randall, J., 1967. "Studies on Neuromelanin. I. A Melanin System in the Human Adult Brainstem," *Neurology,* 17, 512-519.

Bogerts, B., 1981. "A Brainstem Atlas of Catecholaminergic Neurons in Man, Using Melanin as a Natural Marker," *Journal of Comparative Neurology,* 197, 63-80.

Breathnach, A. S., 1988. "Extra-Cutaneous Melanin," *Pigment Cell Research,* 1, 234-237.

Bynum, E. B., 1999. *African Unconscious: Roots of Ancient Mysticism and Modern Psychology.* New York: Teachers College Press.

Chaudhuri, K. R., Hu, M. T., and Brooks, D. J., 2000. "Atypical Parkinsonism in Afro-Caribbean and Indian Origin Immigrants to the UK," *Movement Disorders,* 15, 1, 18-23.

Clark, C., McGee, D.P., Nobles, W., and Weems, L., 1975. "Voodoo or I.Q.: An Introduction to African Psychology," *Journal of Black Psychology,* 1, 2, 9-29.

Commoner, B., Townsend, J., and Pake, G. E., 1954. "Free Radicals in Biological Materials," *Nature,* 174, 689-691.

Cosnett, J. E., and Bill, P. L., 1988. "Parkinson's Disease in Blacks. Observations on Epidemiology in Natal," *South African Medical Journal,* 73, 5, 281-283.

Curzon, G., 1975. "Metals and Melanins in the Extrapyramidal Centers," *Pharmacological Therapeutics Bulletin,* 1, 4, 673-684.

Dahlstrom, A., and Fuxe, K., 1964. "Evidence for the Existence of Monoamine-Containing Neurons in the Central Nervous System I. Demonstration of Monoamines in the Cell Bodies of Brainstem Neurons," *Acta Physiologica Scandinavica,* 62, Supplement 232, 1-55.

de la Monte, S.M., Hutchins, G. M., and Moore, G. W, 1989. "Racial Differences in the Etiology of Dementia and Frequency of Alzheimer Lesions in the Brain," *Journal of the National Medical Association,* 81, 6, 644-52.

de Montellano, B. R. O., 1993. "Melanin, Afrocentricity, and Pseudoscience," *Yearbook of Physical Anthropology,* 36, 33-58.

Eastham, J. H., Lacro, J.P., and Jeste, D. V., 1996. "Ethnicity Disorders," *Mount Sinai Journal of Medicine,* 63,5-6, 314-319.

Faucheux, B.A., Martin, M.E., Beaumont, C., Hauw, J.J., Agid, Y. and Hirsch, E.C. (2003). Neuromelanin associated redox-active iron is increased in the substantia nigra of patients with Parkinson's disease. Journal of Neurochemistry, 86, 1142-1148.

Fenichel, G. M., and Bazelon, M., 1968."Studies on Neuromelanin. IT. Melanin in the Brainstems of infants and Children. *Neurology,* 18, 817-820.

Flaherty, J. A., and Meagher, R., 1980. "Measuring Racial Bias inInpatient Treatment," *American Journal of Psychiatry,* 127, 679-682.

Foley, J. M., and Baxter, D., 1958. "On the Nature of Pigment Granules in the Cells of the Locus Coeruleus and Substantia Nigra," *Journal of Neuropathology and Experimental Neurology,* 17, 586-598.

Gan, E. V., Haberman, H. F., and Menon, I. A., 1976. "Electron Transfer Properties of Melanin," *Archives in Biochemistry and Biophysics,* 173, 666-672.

Gan, E. V., Lam, K. M., Haberman, H. F., and Menon, I.A., 1977. "Electron Transfer Properties of Melanins." *British Journal of Dermatology,* 96, 25-28.

Garver, D. L., and Sladek, J. R., 1975. "Monoamine Distribution in Primate Brain. I. Catecholamine-Containing Perikarya in the Brainstem of *Macaca speciosa,*" *Journal of Comparative Neurology,* 159, 289-304.

Gilbert, C., 1993. "Low Risk to Certain Diseases in Aging: Role of the Autonomic Nervous System and Calcium Metabolism," *Mechanisms of Ageing and Development,* 70, 1-2, 95-113.

Goldman, S.M., and Tanner, C., 1998. "Etiology of Parkinson's Disease," in J. Jankovic and E. Tolosa, eds., *Parkinson's Disease and Movement Disorders,* 3'd ed., 133-158. Baltimore: Lippincott, Williams and Wilkins.

Graham, D. G., 1978. "Oxidative Pathways for Catecholamines in the Genesis of Neuromelanin and Cytotoxic Quinones," *Molecular Pharmacology,* 14, 633-643.

Graham, D. G., 1979 "On the Origin and Significance of Neuromelanin," *Archives in Pathological Laboratory Medicine,* " 103, 359-362.

Hannaford, C., 1995. *Smart Moves: Why Learning Is Not All in Your Head.* Alexander, NC: Great Ocean Publishers.

Joseph, J., Nadeau, D., and Underwood, A., 2002. *The Color Code: A Revolutionary Eating Plan for Optimum Health.* New York: Hyperion.

Kaplan, H. 1., and Sadock, B. J., 1988. "Synopsis of Psychiatry: Behavioral Sciences Clinical," in *Psychiatry,* 5th ed. Baltimore: Williams and Wilkins.

King, R., 1990. *African Origin of Biological Psychiatry.* Tennessee: Seymour-Smith Inc.

Kurtzke, J. F., and Goldberg, I. D., 1988. "Parkinsonism Death Rates by Race, Sex, and Geography," *Neurology,* 38, 10, 1558-1561.

Lanska, D. J., 1997. "The Geographic Distribution of Parkinson's Disease Mortality in the United States," *Journal of Neurological Science,* 150, 1, 63-70.

Lawson, W.B. 1986. "Racial and Ethnic Factors in Psychiatric Research," *Hospital Community Psychiatry,* Jan., 37, 50-54.

Lawson, W.B. 1996a. "Clinical Issues in the Pharmacotherapy of African Americans," *Pharmacological Bulletin,* 32, 2, 275-281.

Lawson, W. B., 1996b. "The Art and Science of the Psychopharmacotherapy of African Americans," *Mt. Sinai Journal of Medicine,* Oct.-Nov., 63, 301-05.

Le Douarin, N. M., and Kalcheim, C., 1999. *The Neural Crest.* New York: Cambridge University Press.

Lin, K. M., Poland, R. E., and Chien, C. P., 1990. "Ethnicity and Psychopharmacology: Recent Findings and Future Research Directions, in E. Sorel, ed., *Family, Culture and Psychobiology.* New York: Legas.

Mann, D. M.A., and Yates, P. O., 1983. "Possible Role of Neuro melanin in the Pathogenesis of Parkinson's Disease,"*Mechanisms in Age Development,* 21, 193-20.

Marder, K., Logroscino, G.,Alfaro, B., Mejia, H., Halim,A., Louis, E., Cote, L., and Mayeux, R., 1998. "Environmental Risk Factors for Parkinson's Disease in an Urban Multiethnic Community," *Neurology,* 50, 1, 279-281.

Marsden, C. D., 1961. "Pigmentation in the Nucleus Substantia Nigra of Mammals," *Journal of Anatomy,* 95,256-261.

—983. "Neuromelanin and Parkinson's Disease," *Journal of Neural Transmission,* 19, 121-141.

McGinness, J., 1972."Mobility Gaps: A Mechanism for Band Gaps in Melanins," *Science,* 177, 896.

McGinness, J., and Proctor, P., 1973. "The Importance of the Fact That Melanin Is Black," *Journal of Theoretical Biology,* 39, 677-688.

McGinness, J., Corry, P., and Proctor, P., 1974. "Amorphous Semiconductor Switching in Melanins," *Science,* 183,853-855.

Monsonego, A., and Weiner, H. L., 2003. "Immunotherapeutic Approaches to Alzheimer's Disease," *Science,* 302,834-838.

Moore, T. O., 1995. *The Science of Melanin: Dispelling the Myths.* Silver Spring, M.D.: Beckham House Publishers.

Moore, T.O. 2002. *Dark Matters Dark Secrets.* Redan, GA: Zamani Press.

Muthane, U., Yasha, T. C., and Shankar, S. K., 1998. "Low Numbers and No Loss of Melanized Nigral Neurons with Increasing Age in Normal Human Brains from India," *Annals in Neurology,* 43, 283-287.

Olszewski, J., and Baxter, D., 1954. *Cytoarchitecture of the Human Brainstem.* Basel: S. Karger.

Olszewski, J., 1964. *Cytoarchitecture of the Human Brain.* New York: Stem and Bitjelow.

Oppenheimer, S. B., and Lefevre, G., 1984. *Introduction to Embryonic Development.* Massachusetts: Allyn and Bacon.

Reichardt, L. F., and Farinas, I., 1999. "Early Actions of Neurotrophic Factors," in M. Sieber-Blum, ed., *Neurotrophins and the Neural Crest,* 1-27. Boca Raton, FL: CRC Press.

Richards, M., and Chaudhuri, K. R., 1996. "Parkinson's Disease in Populations of African Origin: A Review," *Neuroepidemiology,v* 15, 214-221.

Rodgers, A. D., and Curzon, G., 1975. "Melanin Formation by HumanBrain in Vitro," *Journal of Neurochemistry,* 24, 1123-1129.

Rot, et al., 1995. "Corticosteroid Feedback Resistance in Rats Genetically Selected for Increased Dopamine Responsiveness," *Journal of Neuroendocrinology,* 7, 153-161.

Sapper, C. B., and Petito, C. K., 1982."Correspondence of Melanin Pigmented Neurons in Human Brain withA1-A14 Catecholamine Cell Groups," *Brain,* 105, 87-101.

Singleton, A. B., et al., 2003. "Alpha-Synuclein Locus Triplication Causes Parkinson's Disease," *Science,* 302,841.

Squire, L. R., Bloom, F. E., McConnell, S. K., Roberts, J. L., Spitzer, N.C., and Zigmond, M. J., 2003. *Fundamental Neuroscience,* 2nd ed. San Diego, CA: Academic Press.

Strickland, T. L., Lin, K. M., Fu, P., Anderson, D., and Zheng, Y., 1995."Comparison of Lithium Ratio Between African American and Caucasian Bipolar Patients," *Society of Biological Psychiatry,* 37, 325-330.

Strickland, T. L., and Gray, G., 2000. "Neurobehavioral Disorders and Pharmacologic Intervention," in E. Fletcher-Janzen, T. Strickland, and C.R. Reynolds, *Handbook of Cross-Cultural Neuropsychology,* 361-369. New York: Kluwer Academic/ Plenum Publishers.

Tu, et al., 1998. *Annals in Neurology,* 44, 415.

Van Woert, M.H., Prasad, K. N.,and Borg, D. C., 1967. "Spectroscopic tudies of Substantia Nigra Pigment in Human Subjects," *Journal of Neurochemistry,* 14, 707-716.

Van Woert, M. H., 1968. "DPNH Oxidation by Melanin: Inhibition by Phenothiazines," *Proceedings in Social, Experimental and Biological Medicine,* 129, 165-171.

Wood, A. J., and Zhou, H. H., 1991. "Ethnic Differences in Drug Disposition and Responsiveness," *Clinical Pharmacokinetics,* 20, 1-24.

CHAPTER 2

Neuromelanin:
What Is Its Importance in Neural Tissue?
Ann C. Brown, Ph.D.

OVERVIEW

Melanins are pigmented organic biopolymers found in the biosphere, lithosphere, atmosphere, and the cosmos. In living organisms, it is eumelanin, pheomelanin, and allomelanin. Eumelanin is the polymerization of nitrogenous melanogens; pheomelanin is derived from the polymerization of sulfurated melanogens; and allomelanin is the polymerization of polyphenols. In the brain, the black material is referred to as neuromelanin (melanin of neural tissue).

A midsagittal section through the human brain exposes the circumventricular organs (CVO), so-called because of their topology around the deep cavities in the brain, which all contains neuromelanin in some quantity. Neuromelanin is a macromolecule found in a significant amount in catecholaminergic neurons, especially dopaminergic and noradrenergic neurons, in the human brain. Neuromelanin is a brown/black biopolymer pigment found in membrane-bound vesicles in the human central nervous system, primarily in the substantia nigra and locus coeruleus. Neuromelanin has been shown to act as a reservoir for the storage of heavy metals and antioxidants and the formation and scavenging of free radicals. The storage of heavy metals and organic toxins has been interpreted as a protective function for neuromelanin.

Two dementias of the elderly are Alzheimer's and Parkinson's diseases. Brain researchers have found evidence that a protein called ApoE (apolipoprotein E) within nerve cells in the brains of Alzheimer's patients sets off a cascade of biochemical events that decrease and finally destroy neuronal cells in critical areas of the human brain. Studies have shown that in patients with Alzheimer's disorder, a degeneration of synapses and neuronal death occurs in sections of the brain involved in learning and memory. Clinically, Alzheimer's is manifested as memory loss and eventually cell death.

Parkinson's disease is a slow, age-related neurodegenerative disease that results in selective cell death of the neuromelanin pigment-producing neuronal cells in the mesencephalic substantia nigra pars compacta and the locus coeruleus. Nevertheless, there are some unanswered questions regarding the role of neuromelanin in Parkinson's disease. Nuclear magnetic resonance spectroscopy and X-ray diffraction studies have shown that neuromelanin is a multilayer, three-dimensional structure, synthesized enzymatically from the conversion of tyrosine to L-DOPA, dopamine, and DOPA-quinone via tyrosinase. Neuromelanin then reacts with other macromolecules, such as lipids and proteins, and accumulates with age in the neurons as lipofucsin granules in areas of the brain.

Finally, neuromelanin is important in all brain functions, and Parkinson's and Alzheimer's disorders may be different manifestations of the same condition. It appears that most of the brain disorders, to some degree, depend on the neuromelanin, via dopamine and acetylcholine needed for motor control, to maintain functional health. There may be strong implications of poor nutrition in the etiology of many of the long-term neurological diseases. Other possible neuromelanin contributions to consciousness, memory, spirituality, and healthy behavior are considered.

Definition

Melanin is precipitated as black material. Black materials are present throughout the universe (Nicolaus, 1964, 1965): in the biosphere as eumelanin, pheomelanin, and allomelanin; in the lithosphere as minerals, graphites, and fullerenes; in the atmosphere as pollutants and smoke; in the hydrosphere, seas, lakes, and rivers; and in the cosmos as fullerenes and cosmids (Nicolaus, 1964). Melanin is an amorphous semiconductor because it is always in motion, always changing as a result of its central chemical core by adjusting to various energy levels. Because of this motion and by virtue of low resistance to the flow of electrons (Strezelecka, 1982; Riley, 1997; Bynum, 1999), melanins display consistent semiconductor properties. Therefore, melanin serves as a switch for the flow of electrons to higher or lower levels of energy. The low frequency of charge transfer currency gives melanin the ability to bind with strong affinity to drugs and metals, which ultimately damages neural tissue and overall functions via cell death (Swart, 1992; Barnes, 1999).

The presence of melanin in all living organisms has been fully documented and is suggestive of its involvement in many critical functions (perhaps protective) required by all cells to function and maintain life (Lindquist, 1987). Bynum (1999), on the basis of recent experimental data reported in the literature, suggests that neuromelanin may be a superconductor, that is, it appears to have the capacity to enter into a particular state in which energy is conducted through the system with a high degree of efficiency (Cope, 1971, 1978). Bynum further states that these lines of melanin and neuromelanin conductivity in the fetus are present from the earliest stages of embryogenesis. They occur prior to the first heartbeat. In fact, early in embryogenesis there appear "lines of force" that evolve and later may unfold the forms and templates of human cognitive, emotional, and noetic, or spiritual, experience.

In that light, given its bioluminous capacities, its pervasiveness throughout the biological processes, and its crucial significance from the very earliest stages of human life and experience, including life in the womb of the mother, melanin is the chemical of life, the chemical of what our ancestors called the soul, the transformational doorway through which the energy waves of the Holy Soul, Spirit, and Mind pass to take form as the Holy Body.

Ancient Africans in the classical literature *(Meter Neter;* James, 1954) and the psychology of Kemet (Egypt) (Diop, 1974; Akbar, 1984), viewed all the contents of Amenta, the underworld personal subconscious (Mind), the superconscious (Soul), and the collective unconscious (Spirit) as jet black in color (King, 1990, 2001). Barr (1983) lists established and proposed properties of melanin that reflect its numerous functions, and he documents behavior due to its biophysical and biochemical properties.

Barr further advances a major hypothesis that "melanin [in conjunction with other pigment/pigment-related molecules, such as the ubiquitous isopentenoid polymers] functions as the most significant organizational molecule in living systems through its effective *in vivo* control of the vital and diverse covalent current switching." This is a very important concept as it relates to the ability of melanin to generate some degree of low-level electrical current in living systems that generate the current needed for nerve conductivity.

Given the temperature in brain neural tissue, neuromelanin, as an amorphous semiconductor and superconductor can conceivably transmit electrical current without resistance in such a closed environment as perpetual motion, which scientists call a "macroscopic quantum phenomenon" (Nicolaus, 1997). Therefore, nerve cells, for example, are very specialized to receive, process, and transmit information (consciousness) in perpetual motion throughout the body.

This report advances the hypothesis that neuromelanin (brain melanin) is important in neural tissue to direct and transmit the flow of electrical current via nerve cells (neurons) to all portions of the brain. That flow is tantamount to "consciousness." In effect, neuromelanin *is* consciousness. Neurons that are deficient in this black material will show a deficit in current energy, thus jeopardizing consciousness, intellect, motor activity, and other critical functions. Its presence in all mammalian species and in lower vertebrates has been fully documented and speaks for its importance in living organisms. Our interest in neuromelanin is to express a consolidated theoretical approach as psychiatrists, scientists, and psychologists drawing from the wealth of research, including pathological conditions increasingly diagnosed in populations throughout the world. We realized that most of what is known of these pathologies is based on effects, but we must state them, knowing that the causes are always nebulous. Nevertheless, we are also interested in *cause* as well as *effects*. Real answers to pathologies will not be a reality unless causes are first identified.

Except for the role of neuromelanin in pathological states (effects), such as Parkinson's and Alzheimer's diseases and certain neurological tumors, past and present interests by investigators of the role of dark pigment concentrated in several strategic areas of the brain and elsewhere have all but been ignored. As a matter of fact, except for the politics of skin melanin, little attention has been given to its significance in nerve functions. Why is it present in the cell body of a neuron? Even today, the nomenclature concerned with neuromelanin is evaded in the investigative efforts, for it is assumed to be just a pigmented cytosolic waste product of catecholamine synthesis. However, according to Cotzias (1964): "The neuromelanin granule may be the secret key to the understanding of Parkinsonism. God put the melanin granule in the central nervous system (CNS) for a reason. It must be doing something. Something big etc."

Location and Importance of Neuromelanin

Neuromelanin is a dark pigment that aggregates as granules in the cytoplasm of catecholaminergic neurons in the brainstem (mid-brain, pons, and medulla) and the epithelial layers of the ventricles of the brain of humans, and it has been shown to be different from melanin in skin melanocytes (Mason, 1959; Schraermeyer, 1996; Zecca, 1992). It is particularly high in the substantia nigra pars compacta of the mid-brain and noradrenergic neurons in the locus coeruleus of the fourth ventricle. Neuromelanin tends to accumulate with aging, which links it to certain neurodegenerative diseases (Graham, 1978; Bogerts, 1981). The locus coeruleus neuromelanin content has been correlated with feelings of prudence, watchfulness, attentiveness to terror, panic, fear, impulsivity, carelessness, and recklessness (King, 1994).

Classical pioneering data are available from a number of scientists to explain much of cellular interactions during development. Fertilization results from the melanin-containing female egg (Harsa-King, 1980) and the melanin-containing male spermatozoa (Barr, 1983). The sperm entry leaves an electrical pathway of pigment in the frog oocyte (Rugh, 1977; Robinson, 1979). During embryogenesis, a specialized melanized area, Black Dot, (King, 1990) of ectodermal tissue achieves neuronal potential by contact or "primary induction" from the roof of the archenteron (Spemann, 1938). Neural crest forms from this neuroectoderm and folds and separates from the overlying ectoderm to form a neural tube (future spinal cord).

Neural crest cells are bilaterally paired populations of cells arising in the ectoderm at the margins of the neural tube. These pluripotent crest cells migrate away from the neural tube along many pathways to give rise to diverse cell types and derivatives, including sensory and autonomic ganglia, Schwann cells, chromaffin cells, melanocytes, retina, the retinal-pigmented layer, and many other structures (Weston, 1982; Pavan & Tilghman, 1994). The question becomes, what causes this original homogeneous population of cells to migrate and give rise to these diverse cell types. A number of hypotheses have been proposed.

Our hypothesis is that extracellular matrix receptor proteins and intracellular melanin molecules serve as the guiding impetus that gives clues via cell-to-cell signaling as *organizing molecules* (as in Barr, 1983). Cohen and Konigsberg (1975), using the avian system, developed a technique to study neural crest cell migration patterns by removing the crest cells and placing them in culture and subsequently replicating them to establish clones (single precursor population). They found that three types of clones arose: all pigmented cells, all non-pigmented cells, and a mixed population of cells. Other investigators expanded the migration pattern of crest cells and determined that cells migrate in waves and give rise to specific structures along the neural axis of the embryo (Noden, 1975; Bronner-Fraser and Cohen, 1980). For example, in the 2.5-day chick embryos, melanocytes systematically migrated to the sensory ganglia, sympathetic ganglia, adrenal medulla, metanephric primordia, aortic plexus, and some into the gonads (Bronner and Cohen, 1979).

Neural crest cell detachment occurs in craniocaudal waves at the anterior end of the neural tube. This population is the origin of pigmented nuclei (centers) found throughout the brain. The caudal portion of the neural crest gives rise to melanocytes and glial cells (Catala, et al., 2000).

Brain melanin is known as neuromelanin. When viewed in a midsagittal section of the human brain, these pigmented nuclei are strategically located as circumventricular organs (CVO) around the deep cavities of the brain, called ventricles. The CVO include:

1. area postrema in the fourth ventricle
2. median eminence
3. neurohypophysis (posterior pituitary)

4. organum vasculosum of the lamina terminalis

5. pineal gland (epiphysis cerebri)

6. subfonical organ (SFO)

7. subcommissural organ (SCO)

8. choroid plexus (Ganong, 2000)

In other words, neuromelanin is present in neurons at sites that suggest important functions. These CVO neuromelanin have been mapped throughout the brainstem and diencephalon (Bogerts, 1981; Sapper & Petito, 1982; Cowens, 1986).

Histological studies have shown that neuromelanin granules are located in the cytosol of neurons surrounded by a double membrane (Hirosawa, 1968). The brain is a highly complex organ, and numerous investigators are involved in trying to understand more of its various structures and functions. It is composed of neuronal tissue, which consists of two major cell types: neurons and neuroglia. The neuron is the functional unit of the nervous system. Neuroglia (astrocytes, oligodendrocytes, microglia, and ependymal cells) act as glue that fill up most of the spaces between neurons as well as participate in numerous other brain functions. For example, astrocytes assist in establishing and maintaining the blood-brain barrier.

It has been estimated that there are more than 10 billion neurons in the brain that receive and transmit electrical signals throughout the body. Within the soma (cell body) of a neuron are pigment granules that act like batteries and generate current for impulse transmissions away from the soma via extensions called axons, while dendrites receive input and transmit them toward the cell body.

Neurogenesis

Neurogenesis is defined as the transformation of neuroepithelial cells in committed fully differentiated nerve cells. They are determined embryologically by two factors, sonic hedgehop (SHH) and fibroblast growth factor 8 (FGF 8), that induce the formation of dopamine-producing neurons (Simon, Bhatt, et al., 2003). In spite of the understanding that brain tissue is static and does not regenerate itself in the adult mammalian brain, early scanning electron microscopy results show subependymal (hypendymal) cells sending out axon-like processes and subsequently migrating away from the tissue to the cortical area (Privat and Leblond, 1972).

Data reported by Gould, et al., (1997, 1998, 1999) show neurogenesis (new nerve and glial cell formation) in the adult primate neocortex. They reported that new neurons are added to three areas of the brain required for cognitive function: the prefrontal cortex, inferior temporal cortex, and posterior parietal cortex. However, they appear to diminish by stress (Gould, 1998, 2001). They found no new neurons in the primary sensory area (stria cortex), which is involved in visual input. The subventricular area was found to contribute stem cells that migrate through the internal capsule (white matter) to the neocortex where they "sprout" new axons (new neurons).

Other investigators have shown that many subcortical pigmented nuclei and their neuron circuitry contribute to the complex plasticity of the brain that begins its wiring early during embryonic development and provides new neurons throughout the life of the organism (LaCerra and Bingham, 1998). As evidenced from recent studies using various experimental models such as mice (Synder, et al., 1997; Magavi and Macklis, 2001; Chichung, et al., 2002). A diminished population of dopaminergic neurons in the striatal pathway are said to lead to characteristics of Parkinson's disease. Additional information on neurogenesis is provided (see King in Chapter 4).

There seems to be species differences and age distribution variations of catecholaminergic-containing neurons: the guinea pig and rabbit have none, the rat, cat, and dog some, higher primates more, and *Homo sapiens* highest (Mann and Yates, 1983). The differences (King, 1990) in quantity of neuromelanin may suggest that neuromelanin is not a toxic end product, or just a pigmented waste product or some detoxicating protective mechanism, but rather a major role player in electrical impulse transmission throughout the brain tissue, which may be related to the flow of consciousness.

It has been emphasized (Barr, 1983; Lacy, 1984) that neuromelanin has semiconductive properties such as phonon electron coupling as in the Amorphous Theory. Simply stated, the theory says that during an action potential, neuronal phonons are transmitted through the cell and absorbed by neuromelanin granules. With changes in neuromelanin semiconduction (threshold switching), phonons are produced and transmitted via gap junctions between cells, influencing the polarization and permeability of ions via an all-or-none phenomenon (McGinness, et al., 1974). King (1990) states that since a great majority of catecholamine melanin containing neurons are dopaminergic, in general they may be involved in conscious perception, movements, emotions, and memory. Comparing pigmented versus nonpigmented neurons, pigmented neurons could process information differently or more efficiently (McGinness, 1985).

Neurotoxin MPTP and Its Effect on Neuromelanin

It has been suggested that neuromelanin may play a critical role in protecting neurons from high ionizing radiation and other harmful things such as free radicals, heavy metals, and heroin-like neurotoxic agents. Such a neurotoxic agent, MPTP (1-methyl-4-phenyl-1,2,3,6-tetrahydropyridine), selectively destroys substantia nigra neurons due to the accumulation of its toxic metabolite, methylphenylpyridine (MPP+), in melanin granules of primates

(D'Amato et al., 1986; Lindquist, 1987; Levi et al., 1989; Zecca, et al. 2001). When the synthetic MPTP is administered to humans or primates, they develop Parkinson-like symptoms (Barbeau, 1985). The active form, MPP+, is thought to poison the electron transport system (ETS) complex I and destroys mitochondria and selectively destroys dopamine-producing neurons of the substantia nigra (Yantiri and Anderson, 1999).

These investigators suggested that iron and MPTP may work cooperatively to evoke deleterious effects by depleting cells of free radical protections (superoxide dismutase (SOD) and glutathione peroxidase) since these enzymes were protective in transgenic and knockout mouse studies. Therefore, MPP+ accumulates via neuromelanin uptake causing toxicity and neuron cell death by inhibition to mitochondrial ATP production via NADH dehydrogenase and coenzyme Q depletion and loss of glutathione and calcium (Barbeau, 1986; Jenner, 1989).

In addition, MPP+ mediates oxidative stress and neurotoxicity via the dopamine pathway, catalyzed by iron in the substantia nigra pars compacta. High levels of iron seen in Parkinson's patients and the neurotoxin MPTP can cause accumulative neurodegeneration of dopaminergic neurons, depleting the cells of neuromelanin required for impulse generation (Larsson, 1993). Iron-induced free radical formation has been implicated in Parkinson's disease (Beard, et al., 1993).

Neuromelanin in the Brain Cavities

The deep cavities of the cortex and its cells are discussed separately to show connections and that their lining cells contain pigmented cells. These deep cavities are part of the phylogenetic ancient brain. During embryogenesis, neuromelanin with its alternating dark and light band gaps on the UV-visible spectrum have the ability to absorb photons of light that are not re-radiated out into the system and, therefore, facilitate in organizing deeper, complex biological and developmental structures (Bynum, 1999).

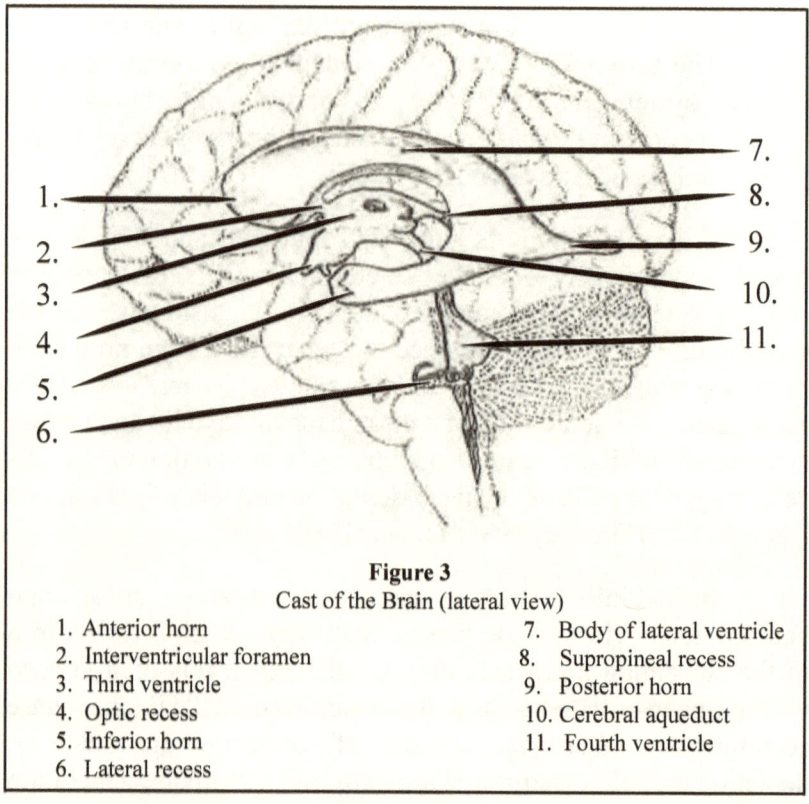

Figure 3
Cast of the Brain (lateral view)

1. Anterior horn
2. Interventricular foramen
3. Third ventricle
4. Optic recess
5. Inferior horn
6. Lateral recess
7. Body of lateral ventricle
8. Supropineal recess
9. Posterior horn
10. Cerebral aqueduct
11. Fourth ventricle

Figure 3 is a cast of the deep cavities (ventricles) found in the cerebrum that shows the communication of the four cavities (ventricles) in the developed brain. Two ventricles are located within the cerebral hemispheres (lateral ventricles) and a third ventricle/choroid plexus in the upper brainstem of the diencephalon. A fourth ventricle/choroid plexus is located between the cerebellum and the brainstem. The choroid plexus is believed to secrete most the cerebrospinal fluid. The lateral ventricles are closed cavities except where they communicate with the third ventricle via two interventricular foramina (foramina of Monro).

The fourth ventricle communicates with the third ventricle of the pons and medulla through the cerebral aqueduct (aqueduct of Sylvius) of the mid-brain and by three apertures: the unpaired median aperture (foramen of Magendie) and two lateral apertures (foramina of Luschka) that direct the cerebrospinal fluid into the subarachnoid space. The fourth ventricle contains highly pigmented centers called the locus coeruleus (Scott, et al., 1973). The fourth ventricle continues caudally as a narrow central canal through the gray matter of the spinal cord.

All four cavities of the ventricular system are lined with an epithelium of ependymal cells that secrete cerebrospinal fluid involved in many brain functions. Deep in the cerebral hemisphere (subcortical areas) are masses of nuclei or centers of gray matter called basal ganglia (caudate nucleus, putamen, globus pallidus, claustrum, and amygdala) that also contain neuromelanin (Akert, 1969; Smith, 1970).

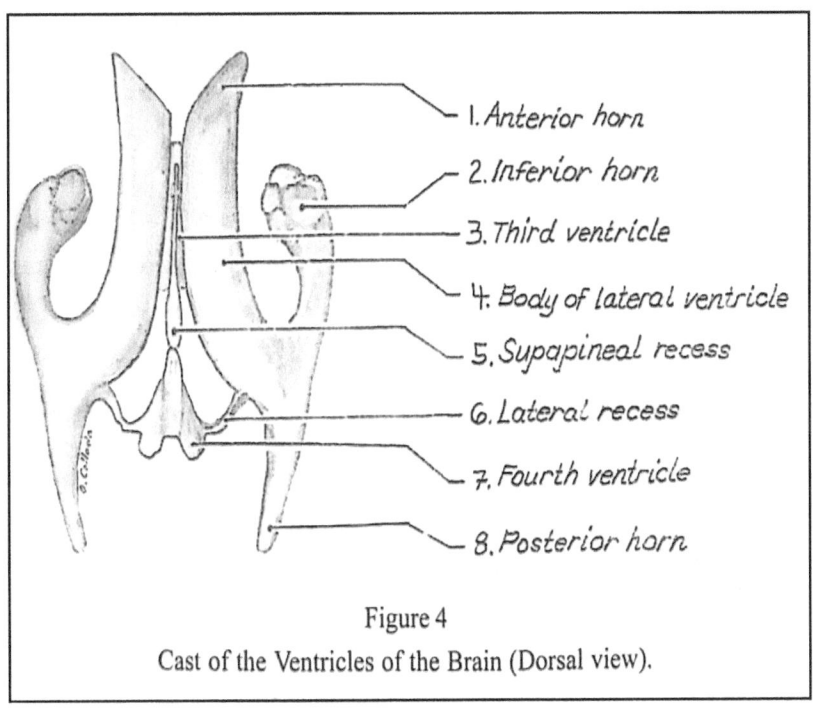

Figure 4
Cast of the Ventricles of the Brain (Dorsal view).

Figure 4 is a dorsal view of these ventricular cavities, showing their communicative connections. Cerebrospinal fluid bathes these cavities. These masses form a portion of the ventricular wall, separated by white (myelinated) fibers. Again, these pigmented nuclei have been shown to be critical in many functions such as motor (skeletal muscle) activity. Data have been presented to show that, collectively, the pigmented cells are responsible for the programming, inception, and termination of motor movement. The symptoms observed in Parkinsonism are consistent with the degeneracy observed in these areas.

Biosynthesis of Neuromelanin

The steps involved in the synthesis of neuromelanin are still under debate. Some investigators argue that its synthesis is from a series of enzymatic steps similar to those established in eurmelanin (brown-black melanin), and others argue that its presence in the substantia pars compacta is from autoxidation of dopamine derivatives (Odh, et al., 1994; Zecca, et al., 2001, 2002).

Recently, Tief and Beerman (1998) reported on melanin synthesis of proteins from extracts of adult mouse brain. They detected tyrosinase promoter gene activity in the cerebral cortex, olfactory system, hippocampus, epithalamus (pineal), and substantia nigra pars compacta areas that corresponded to positive staining during embryogenesis. They further stated that tyrosinase and tyrosinase-related proteins (TRP-1 and TRP-2) have been shown by others to be expressed in neural crest-derived melanocytes and retinal-pigmented epithelium. They concluded that tyrosinase promoter activity is active throughout murine brain development and could be the enzymes required for neuromelanin biosynthesis.

Strong evidence of the synthesis of neuromelanin has been documented by the Sulzer (2000) team. Experimentally, they induced neuromelanin in rat ventral mid-brain substantia nigra and cultured PC-12 cell line with L-DOPA (an intermediate in the neuromelanin pathway). Using electron paramagnetic resonance (EPR), they demonstrated that L-DOPA was converted to dopamine in the cytosol and that the newly converted pigmented granules were localized in double membrane vesicles identical to those observed in the substantia nigra pars compacta of humans. These researchers were able to abolish neuromelanin synthesis by adenovirus constructs and concluded that neuromelanin results from excess catecholamine not in vesicles in the cytosol and subsequently stored in vesicles.

The controversy of the biosynthesis continues. Is it similar to skin melanin? Is tyrosine hydroxylase the rate-limiting enzyme in the dopamine/neuromelanin pathways or is it tyrosinase? Whatever its origin, neuromelanin presence in critical areas of brain/eye/ear functions is evident by its decrease in cell death seen in Parkinsonism and Alzheimer's disorder and other neurodegenerative diseases. With advances in brain research technology, instruments with strong sensitivity such as MRI and immunocytochemistry techniques, it is now possible to probe the deeper areas of the brain for clues of tissue damage and cell death. With such probe technology, is it then possible to determine if the loss of neurons manifested as degeneracy or neuromelanin loss in such neurons and other brain cells cannot function in a normal manner? Recently scientists were able to probe the memory-related structures of the brain for cellular degeneration in cases of Parkinsonism and Alzheimer's diseases (Shu and Wu, et al., 2002). Their findings have consistently pointed to a decrease in neuromelanin-bearing nerve cells.

Relationship of Neuromelanin in Parkinson's and Alzheimer's Disorders

One of the oldest neurodegenerative dementias observed clinically in humans is Parkinson's disease. It is associated with defective catecholaminergic neurons and neural circuitry. Specifically, there is a deficit in dopamine (neuromelanin) of the mid-brain neurons. The neurotransmitter dopamine as well as acetylcholine plays a major role in skeletal muscle movement and other functions. A representative summary of this widely studied dementia by the National Institute of Environmental Health Sciences, 1999, and *Science-Week Focus Report* make the following points: a) The disease affects more than 1 million people in North America and age is a consistent risk factor; (b) Parkinson's disease occurs throughout the world in all ethnic groups, mainly after 50 years of age. Its lowest prevalence is among Asians and African Blacks and its highest is among Whites; (c) Clinical symptoms include tremor, rigidity, slurred speech, and uncontrolled voluntary motor movements; (d) There is a progressive cell death of the dopaminergic pigmented nuclei and neural circuitry in the substantia nigra (mid-brain region), as well as other subcortical pigmented nuclei that are responsive to the complex plasticity of normal brain functions; (e) The mechanisms responsible for cell death in Parkinson's disease are said to be unknown. Some of the mechanisms include genetic factors, increasing age, environmental factors, immune factors, free radical toxicity, and nutritional (dietary) factors. In some Parkinson's patients with abnormal intracellular accumulation of protein inclusions called alpha-synuclein, components of Lewy bodies have been observed in the substantia nigra (Marsden, 1983; Ballard, et al., 1998; Double, 1999). Treatment is based on clinical symptoms and attempts to restore some of the intermediates in the neuromelanin pathway, L-DOPA (3,4-dihydroxyphenylalanine) to be converted to dopamine, for example. Numerous labs are at work examining neuromelanin containing neuron populations in tissue (cell) culture in the presence of L-DOPA (Sulzer, 2000) in an attempt to understand the relationship of these neurons and their role in the disease.

Neuromelanin: What Is Its Importance in Neural Tissue?

There is not necessarily an age limit for Parkinson's disease. For example, Pope John Paul II showed symptoms of Parkinson's disease in which his left arm shook uncontrollably at rest while his head tilted sharply to the right. Other Parkinson's patients that show similar symptoms include ex-boxer Muhammad Ali, Michael J. Fox, and former U.S. Attorney General Janet Reno. We feel that neuromelanin plays a major role in the formation and transmission of neurotransmitter substances from one neuron to others as signals for consciousness and intellect to be facilitated.

Alzheimer's disease is the most prevalent neurodegenerative disorder in humans. It involves a loss of normal capacity to reason, think, recognize, and function. This means that certain brain cells and neurons are damaged, followed by cell death due to the accumulation of neurotoxins in neurons (Cafe, et al., 1996). Two pathologies that are hallmarks of this disorder are abnormal clumps (now called amyloid plaques) and tangled bundles of fibers (presently called neurofibrillary tangles of axons and dendrites, Masliah, 2000). Again, age is an important risk factor. It has been observed from imaging techniques seen in Alzheimer's patients that there is a calculated loss of nerve cells, which affects memory and language. Specifically, cell loss and cell death have been demonstrated in the hippocampus and the cortical area that deals with the storage (learning) and retrieval (memory) of information (Gage and Eriksson, 1998). In other neurons abnormal deposits of proteins, called alpha-synuclein and ubiquitin, within the vesicles with Lewy bodies create "sticky" plaques, causing the nerve axons to retract, twist, curl, and coil, thus preventing the transport of neurotransmitter release from the axon and jeopardizing nerve transmission (Sulzer, 2001).

Many Alzheimer patients tend to be forgetful, especially of recent events. As the disease progresses, the patient becomes confused and may forget where he/she resides. Later, they may have problems speaking, reading, writing, and understanding and tend to wander away from home. The causes of Alzheimer's disease have not been fully established. Implications for nerve damage are similar to those of Parkinson's disease and may, in fact, be a different manifestation of a similar disorder, that is, a loss of neuromelanin needed for impulse formation. Again, the contributing causes are related to neurotoxins, inadequate nutrient molecules to regenerate new neurons below a critical level to effectuate electrical energy transmission for consciousness, and other functions. These pathologies not only affect motor skills but behavioral functions as well (Moore, 2002).

Prevalence data show that about 50,000 new cases showing symptoms are diagnosed in the United States each year. In 1989, this prompted then President George Bush to sign into law House Joint Resolution 174, declaring the 1990s as the "Decade of the Brain," and making available more research funds for neuroscience research. Recent studies show that African Americans and Asians are less likely to exhibit Parkinson symptoms than people of European descent, even though some investigators have concluded that Asians and African Blacks have a lower incidence compared to American Blacks and Whites (Lang, et al., 1998).

In light of this discussion, the differences in prevalence of neurodegenerative diseases in the major races may be due to neuromelanin biochemistry and nutritional intake in different cultures. There is a need to reexamine the types of foods that can resupply the cells throughout the body with those pigments found in food as raw materials for nerve regeneration and maintenance.

The cost of Parkinson's disease in the United States is estimated to surpass $5.6 billion annually. The average patient spends about $2,500 annually on medication. This cost has put an economic strain on the geriatric population with fixed income who are already on many other medications. Many of these patients are unable to financially manage costs that exceed their medical coverage. The African American Parkinson's patient is less able to pay the $25,000 for the overall surgery required to install electrodes needed to reduce tremors. Therefore, healthcare in the elderly is prohibitively costly.

THE ROLE OF FREE RADICALS AND FREE RADICAL SCAVENGERS IN PARKINSON'S DISEASE

A free radical (FR) is a chemical species capable of independent existence that contains one or more unpaired electrons that occupy an atomic or molecular orbit by itself (Halliwell and Gutteridge, 1984). Such unpaired electrons cause a chemical species to be paramagnetic (attracted slightly to a magnetic field) and thus highly reactive. Free radicals are unstable due to the existence of at least one unpaired electron in its outer orbital. It is the pairing of the electrons that renders them stable. Unpaired electrons have a tendency to form a chemical reaction with other chemical species and create a potential danger, which can cause harm to the cellular mechanisms, especially those involving oxygen ions (Gerschman, 1959).

The oxygen-free radicals, or ROS (reactive oxygen species), include: superoxide anion, hydroxyl radicals, lipid peroxyl radical, singlet oxygen, hydrogen peroxide, and hypochlorous acid. Most of the transition metals contain unpaired electrons and can qualify as radicals. This is especially true of iron (Fe). Iron (Fe^{+2}) and hydrogen peroxide (H_2O_2) can react with many organic molecules such as those that occur in neural tissue. This reaction has become known as the Fenton Reaction (Fenton, 1894). When combined, the superoxide anion and hydrogen peroxide can be scavenged in the presence of the transition metal, iron, acting as a catalyst for decreasing FRs in a combined reaction known as the iron-catalyzed Haber-Weiss Reaction (Haber and Weiss, 1934). The reaction is summarized as

$$O_2^- + H_2O_2 \xrightarrow{\text{Fe-salt}} O_2 + OH^* + OH^-$$

(Superoxide anion) + (Hydrogen peroxide) catalyst (Oxygen) + (Hydroxyl radical) + (Hydroxyl anion)

Cells have multiple indigenous protective mechanisms against FR damage in the form of free radical scavengers (FRS). Many of these protective molecules are classified as antioxidants, for example: Vitamin C (ascorbic acid), beta-carotene, vitamin E (tocopherols), glutathione, and enzymes such as superoxide dismutase (SOD), catalase, and organic peroxidases. Experimentally, it has been demonstrated that normal cellular functions may become disturbed or altered when an abnormal balance of FRs or FRSs are present in the cellular environment (Brown and Lutton, 1988; Kensler and Taffe, 1986). Glutathione has been shown to be a major FRS tripeptide found in most mammalian cells in that it facilitates the destruction of quasi-stable hydrogen peroxide and other organic peroxides, which can generate toxic FR species (Kuthman and Eriksson, 1979). The Glutathione Cycle used by the cell to scavenge FR species plays a critical role in the detoxification of ROS (Brown, 2003, minireview).

The major function of the FRS glutathione is to protect the cell against endogenous FRs and other oxygen stressors. In neurodegenerative diseases such as Parkinsonism and Alzheimer's, FRs are thought to be produced within the deep cerebral nuclei called basal nuclei (striatonigral), and this can lead to progressive damage and nigral death (Zeevalk, Bernard, and Ehrhart, 2003). Recent evidence suggests that membrane lipids in the substantia nigra show typical signs of oxidative damage, suggesting FR injury via lipid peroxidation (incorporation of ROS in the membrane lipid moiety of neuronal cells) (Hirsch, 1993; Jenner and Olanow, 1996; Simonian and Coyle, 1996; Sudha, Rao, and Rao, 2003 Faucheux and Martin et al., 2003). Studies of glutathione (GSH) depletion in vitro and in vivo in the presence of the neurotoxin MPTP increased ROS in mice, suggesting potential damage to mid-brain neurons (Sriram, Pai, K.S., et al., 1997).

The debate continues as to the exact etiology of neurological disorders in patients with Parkinson's and Alzheimer's diseases. It is theoretically possible that oxidative stress can occur in neuronal tissue because of an imbalance in FR/FRS production and the ability of the neurons in the brain to protect and prevent cytotoxic radical formations leading to a progressive, long-term deterioration of motor functions and cognition.

Although, a scarcity of scientific research is available of the role of nutrition as an undying *cause* of neurodegenerative disorders, poor nutritional habits over numerous years may contribute to many brain disorders. For example, high consumption of meats and foods with metal toxicity can metabolically generate ROS that can damage normal tissue, especially when antioxidant glutathione and other free radical scavengers are reduced. A daily intake of foods such as fruits, vegetables, nuts, and seeds (especially when consumed raw) and regular exercises contribute molecules that are supportive of good health and wellness. Nutrition is an area that needs to be thoroughly examined and documented. Poor nutrition may be one of the main contributing *causes* of Parkinson's and Alzheimer's diseases. In addition, there is some hint that years of negative thinking, including fear, resentment, worry, bitterness, and anger, as well as body toxic load may have a strong effect on overall health. Nerve cells are certainly influenced by the biochemical milieu of the blood stream.

Natural Plant Pigments in Foods Are Required for Normal Nerve Functions

At the beginning of this work, the ubiquity of black material in the universe was well documented. Investigators have shown the presence and importance of this dark pigment (melanin/neuromelanin) in the cytoplasm of trillions of nerve cells in the brain. It was advanced that neuromelanin is concentrated in neurons and cells in strategic areas of the brain, which is key to maintaining the flow of impulses throughout the brain continuously, a process we call *consciousness*. This very powerful black matter acts as a low-level semiconductor (superconductors), transmitting and regenerating the current needed for consciousness, intellect, and sensory and motor functions. It is our position that neuromelanin is regenerated through raw food and food pigments intake, especially from raw fruits and vegetables and natural unprocessed food additives.

Much attention has been focused on Alzheimer's disease, an insidious neurodegenerative brain disorder that results in memory loss, unusual behavior, and regressive thinking ability. These neurological patterns are said to be related to the death and destruction of brain nerve cells, connective tissue, and supporting cells called neuroglia. The preceding sections describe neuromelanin deficiency in Alzheimer's and Parkinson's brain disorders. This section advances the notion that oxygen and natural (not synthetic) food pigments can prevent the coagulation of toxic protein in the nerve cell body and neurofibrillary abnormal twisting that results in memory loss due to the inability of nerve impulses to be propagated (review, UCLA Alzheimer's Disease Research Center).

For more than three decades scientists have thoroughly documented the protective effect of one natural food ingredient: curcumin, a yellow, pigmented chemical constituent derived from turmeric and other spices used in foods (Lin, et al., 1994). Over 100 Western studies, and still growing, and many more by Indian scientists have already demonstrated that natural plant-derived phytochemical, polyphenolic pigments in common food additives ingested daily in cultures such as Indian (Ayurvedic), African, Caribbean, Asian, and South American can prevent or reduce many of the neurological and other debilitating diseases seen in Western cultures where these additives are rarely used. In particular, studies show that low doses of curcumin (*Curcuma longa*), a yellow, active ingredient in curry and turmeric, may reduce or prevent conditions ranging from Alzheimer's to certain viral diseases (Aggarwal, Kumar, and Bharti, 2003). This would suggest that food pigments are biochemical molecules that are as important as vitamins, minerals, and other food components and must be included in daily consumption for healthy nerve functioning.

Studies done on the elderly Indian population that regularly consumes large amounts of turmeric spice in their diet shows them less likely to develop Alzheimer's or multiple sclerosis (MS) than their counterparts in the Western population (Natarajan and Bright, 2002). In view of this, Western scientists theorize that curcumin must contain anti-inflammatory properties. This conjecture was supported by a finding that Westerners taking anti-inflammatories regularly for arthritis are less likely to develop Alzheimer's disease (Halliday, Robinson, and Kril, 2000).

Recent studies theorize that curcumin may effectively cross the blood-brain barrier and bind to toxic beta amyloid proteins in the brains of Alzheimer's patients to breakup existing plaques and prevent the formation of others (Yang, et al., 2004). Furthermore, curcumin has powerful anti-oxidative and anti-inflammatory properties that in low doses ease disease symptoms caused by oxidation and inflammation (Huang, et al., 2004; Kelawata and Ananthanarayan, 2004; Yang, 2004; Frautschy, 2001). Furthermore, curcumin may be chemoprotective against the growth of gastric and colon cancer (Mahady, et al., 2002) and may protect the brain against free radical damage by the induction of heme oxygenase as protection (Scapagnini, et al., 2004).

It is our position that a great deal of the pain and suffering, visited upon the world's elderly population especially, are nutrition related. In spite of food distribution injustices, poor countries often have advantages over more affluent countries in their cultural selection of foods that tend to keep certain diseases at a minimum.

We further hypothesize that pigment-bearing raw food substances make available biochemical molecules used by nerve cells to regenerate neuromelanin that the cell bodies of millions of neurons require moment-by-moment as semiconductors (superconductors) and for the propagation of impulses that we call consciousness. Any obstructions of this current flow by foreign toxic matter, such as toxins from years of waste in the colon and recirculation to the brain, affect the neuromelanin and may contribute to Alzheimer's pathology, as seen in hippocampus brain tissue, or the rigidity and tremors due to lack of neurons and their neuromelanin in the mid-brains of Parkinson's patients. It is the author's observation that at the site of a synapse, there is neuromelanin that receives and recharges the impulse. A second look at neuromelanin as a semiconductor in neurons needs to be carefully taken to find answers to age-old brain disorders. More on nutrition in this disease etiology is forthcoming in a separate work.

CONCLUSION

Melanin is a biopolymeric pigment with semiconductor properties. It is present within the cells of all living things. Since these pigments are present from the beginning of conception, these bioactive pigments play a major role in the cellular/molecular and embryological organization of all living things. Neuromelanin is present in neural tissue and appears to function in the transduction and flow of electrical current from one part of the brain to another through a complex circuitry. In brain tissue, data show that neuromelanin located within dopaminergic and adrenergic neurons may exhibit properties that qualify it as both a biological superconductor and an amorphous semiconductor, in other words a threshold switch similar to that of inorganic materials (McGinness, et al., 1974).

McGinness in 2001 reports: "Melanins give a flash of light when they switch-clearly electroluminescence, though its significance is not completely understood at this time." Is this flash of light what is called biophotons? If so, melanin may be the source of this light. Does this flash contribute to what is called spirituality, the ability to utilize brain energy for consciousness and healing qualities? In the brain, could this flash of light generated by such amorphous semiconductivity between low to high threshold switching of neuromelanin be analogous to "consciousness"? If this is the case, then this neural dark matter also is the light that makes it possible to "see." And if neuromelanin pigment is stored within vesicles, its synthesis may be required as a continuous source of current whose product is the neurotransmitter substance. Some of these ideas need to be discussed and, where possible, investigated from a new perspective to gain additional answers to health issues.

In the context of health-related research we have also noted that natural pigments found in foods and spices are beneficial to the regeneration of neuromelanin found in the body, especially in the brain where the pigment is key as a semiconductor that propagates electrical currents continuously via trillions of neurons, which allows consciousness and other functions of life. Elderly persons in cultures where spices such as turmeric (curcumin) are a regular food additive rarely suffer debilitating diseases such as arthritis, Alzheimer's, and certain other neurodegenerative diseases. Turmeric (curcumin) demonstrates anti-oxidative, anti-inflammatory, and cytoprotective properties by acting as a scavenger of reactive oxygen species that may contribute to the formation of beta-amyloid plaques, which sequester neuronal cytoplasmic neuromelanin and other toxic chemicals, causing some of the symptoms seen in nervous system diseases.

In addition, the neuromelanin of the neuronal-neuroglia ventricular system may activate and then transmit consciousness to all parts of the brain for cognition and motor activity. Data clearly show that a decrease in the mid-brain dopamine-bearing neuron/neuroglia population results in a loss of motor activity, speech, learning, and memory as seen clinically in Parkinson's and Alzheimer's disorders (Kastner, et al., 1992). In some patients, the axons and dendrites of neurons in certain areas of the brain are twisted and coiled, preventing neurotransmitter flow from one nerve to another, which contributes to pathologies and cell death.

The specific causes of these neurodegenerative disorders have not been clearly established. There appear to be many factors such as genetic inheritance, the environment, increasing age, free radical toxicity, neurotoxins, and nutrition. Drugs exert cumulative negative effects, especially in people of color, that contribute to behavioral modifications. The importance of these effects needs to be taught to our children, especially nutritional considerations. Yes, darkness matters (Moore, 1995, 2002) and our scientific perspective should be expanded and explored to include an afrocentric perspective.

Nevertheless, there is some hope in addressing some of the disease symptoms originating in areas of the brain such as the hippocampus and the basal ganglia (dentate gyrus), where experiments with adult monkeys have shown stem cells to "sprout" into new neurons to replace damaged cells (Gould, et al., 1997, 1998, 1999, 2002). Finally, we would encourage some open, honest scholarly discussion on the meaning and importance of neural dark matter as it relates to behavior, emotions, learning, intelligence, health, well-being, aging, and longevity under non-pathological conditions.

REFERENCES

Aggarwal, B. B., Kumar, A., and Bharti, A. C., 2003. "Anticancer Potential of Curcumin: Preclinical and Clinical Studies," *Anticancer Res.,* 23, 368-98.

Akbar, N., 1984. *Chains and Images of Psychological Slavery.* Jersey City, NJ: New Mind Productions.

Akert, K., 1969. "The Mammalian Subfornical Organs," *J. Neuro Visceral Rel.,* Suppl. IX, 78-93.

Ballard, C., Grace, J., and Holmes, C., 1998. "Neuroleptic Sensitivity in Dementia with Lewy Bodies and Alzheimer's Disease, *Lancet,* 351, 1032-1033.

Barbeau, A., 1985. "Comparative Behavioral Biochemical and Pigmentary Effects of MPTP, MPP+ and Paraquat," in *Rana pipiens, Life Sci.,* 37, 1529-1538.

-1986. "Environmental and Genetic Factors in the Etiology of Parkinson's Disease." *Adv. Neural.,* 45, 299-306.

Barnes, C., 1988. *Melanin: The Chemical Key to Black Greatness,* vol. 1, 56, 57, "Black Greatness" series. Houston, TX.

Barr, F., 1983. "Melanin: The Organizing Molecule," in *Medical Hypotheses,* 11, 1-140, review.

Beard, J. L., Connor, J. D., and Jones, B. C., 1993. "Brain Iron: Location and Function," *Prog. Food Nutr. Sci.,* 17, 183-221, review.

Bogerts, B., 1981. "A Brainstem Atlas of Catecholaminergic Neurons in Man, Using Melanin as Natural Marker," *J* Comp. Neurol., 197, 63-80.

Bonner, M. E., and Cohen, A. M., 1979. "Migratory Patterns of Cloned Neural Crest Melanocytes Injected into Host Chicken Embryos," *Proc. Natl. Acad. Sci. US. A.,* 76, 1843-1847.

Bonner Fraser, M., and Cohen, A. M., 1980. "The Neural Crest: What Can It Tell Us About Cell Migration and Determination?" *Curr. Top. Dev. Biol.,* 15 Pt. 1, 1-25, review.

Brown, A. C., 2003. "Glutathione: Free Radical Scavenger That Protects Against Cell Damage," *In Vivo,* 24, 4-10, minireview.

Brown, A. C., and Lutton, J. D., 1988. "The Significance of Free Radicals and Free Radical Scavengers," *Advances in Experimental Medicine and Biology,* 241, 135-148.

Bynum, E. B., 1999. *The African Unconscious. Roots of Ancient Mysticism and Modern Psychology.* NY, NY: Cosimo Books.

Cafe, C., Torri, C., Bertorelli, L., et al., 1996. "Oxidative Stress After Acute and Chronic Application of Beta Amyloid Fragments 25-35 in Cortical Cultures," *Neuroscience Lett.,* 203, 61-65.

Catala, M., Ziller, C., et al., 2000. "The Developmental Potentials of the Caudal most Part of the Neural Crest Are Restricted to Melanocytes and Glia," *Mech. Dev.,* 95, 77- 87.

Chichung, Lie, D. G., Dziewczapolski, A. R., Willhopite, et al., 2002. "The Adult Substantia Nigra Contains Progenitor Cells with Neurogenic Potential," *J Neurosci.,* 22, 6639-6649.

Cohen, A. M., and Konigsberg, L. R., 1975. "A Clonal Approach to the Problem of Neural Crest Determination," *Dev. Biol.,* 46, 262-280.

Cope, F. W., 1978. "Discontinuous Magnetic Field Effects (Barkhausen noise) in Nucleic Acids As Evidence for Room Temperature Organic Superconductors, "*Physiological Chemistry and Physics,* 10, 233-245.

Cope, F. W., 1981. "Organic Superconductive Phenomena At Room Temperature. Some Magnetic Properties of Dyes...," *PhysiologicalChemistry and Physics,* 13, 99-110.

Cotzias, G. C., et al., 1964. "Melanogenesis and Extrapyramidal Diseases.," *Fed Proc.,* 23, 713-718.

Cowens, D., 1986. "The Melanosomes in the Human Cerebellum (nucleus pigmentosis cerebellaris) and Homogules in the Monkey," *J Neuropath. Exper. Neural.,* 45, 205-221.

D'Amato, R. J., Lipman, Z. P., and Snyder, S. H., 1986. "Selectivity of the Parkinsonian Neurotoxin MPTP: Toxic Metabolite MPP+ Binds to Neuromelanin," *Science,* 231, 987-989.

Diop, A. C., 1991. *Civilization or Barbarism: An Authentic Anthropology.* Brooklyn, NY: Lawrence Hill Books.

Double, K. L., Riederer, P., and Gerlach, M., 1999. "Significance of Neuromelanin for Neurodegeneration in Parkinson's Disease," *Drug News & Perspectives,* 12, 6.

Double, K. L., Zecca, L., et al., 2000. "Structural Characteristics of Human Substantia Nigra Neuromelanin and Synthetic Dopamine Melanins," *J Neurochem,* 75, 2583-2589.

Dzierzega-Lecznar, A., Kurkiewics, S., et al., 2004. "GC/MS Analysis of Thermally Degraded Neuromelanin from the Human Substantia Nigra," *J Am. Soc. Mass Spectrom.* 15(6), 920-926.

Faucheux, B.A., Martin, M-E., Beaumont, C, et al., 2003. ''Neuromelanin associated redox-active iron is increased in the substantia nigra of patients with Parkinson's disease,' J. Neurochem. 86, 1142-1148.

Gage, F., and Eriksson, P., 1998. "Neurogenesis in the Adult Human Hippocampus," *Nature Med.,* 4, 1313-1317.

Ganong, W. F., 2000. "Circumventricular Organs: Definition and Role in the Regulation of Endocrine and Autonomic Function," *Clin. Pharmacol. Physiol.*, 27, 422-427.

Gerschman, R., 1959. "Oxygen Effects in Biological Systems," *Proc. Int. Congr. Physiol. Sci.*, 222-226, 21st Buenos Aires.

Gould, E., et al., 1997. "Neurogenesis in the Dentate Gyms of the Adult Tree Shrew Is Regulated by Psychosocial Stress andN.M.D.A Receptor Activation," *J. Neurosci.* 17, 2492-2498.

Gould, E., Tanapat, E., McEwen, B. S., et al., 1998. "Proliferation of Granule Cell Precursors in the Dentate Gyrus in Adult Monkeys is Diminished by Stress," *Proc. Natl. Acad. Sci. U.S. A.*, 95,31689-31710.

Gould, E., Beylin, A., Tenapat, P., et al., 1999. "Learning Enhances Adult Neurogenesis in the Hippocampal Formation," *Nature Neurosci.*, 2, 260-265.

Gould, E., Vail, N., Wagers, M., and Gross, C. G., 2001. "Adult Generated Hippocampal and Neocortical Neurons in Macaques Have a Transient Existence." *Proc. Natl. Acad. Sci. U. S. A.*, 98, 10910-10917.

Gould, E., and Gross, C. G., 2002. "Neurogenesis in Adult Mammals:Some Progress and Problems," *J. Neurosci.*, 22, 619-623.

Graham, D. G., 1978. "Oxidative Pathways for Catecholamines in the Genesis of Neuromelanin and Cytotoxic Quinones," *Mol. Pharmacol.*, 14, 633-643.

Halliday, G., Robinson, S. R., Shepherd, C., and Kril, J., 2000. "Alzheimer's Disease and Inflammation: A Review of Cellular and Therapeutic Mechanisms," *Clin. Exp. Pharmacol. Physiol.*,

27, 1-8, review.

Halliwell, B., and Gutteridge, J.M. C., 1984. "Oxygen Toxicity, Oxygen Radicals, Transition Metals and Disease," *Biochem. J,* 219, 1-4.

Harsa-King, M., 1980. "Melanogenesis in Oocytes of Wild Type and Mutant Albino Axolotls. *Dev. Biol.,* 74, 251-262.

Hendrie, H., Ogunniyi, A., Hall, K. S., et al., 2001. "African Americans Develop Alzheimer's Disease and Other Dementias At Twice the Rate of Africans," *JAMA,* 285, 739-747.

Hirosawa, K., 1968. "Electron Microscopic Studies on Pigment Granules in the Substantia Nigra and Locus Coeruleus of the Japanese Monkey *(Macaca fascularis yakui),"* Z. *Zellforsch Anat.,* 88, 187-203.

Hirsch, E. C., 1993. "Does Oxidative Stress Participate in Nerve Cell Death in Parkinson's Disease?" *Neurology,* 33 (suppl. 1), 52-59.

Huang, X., Moir, R. D., Tanzi, R. E., Bush, A. L., and Rogers, J. T., 2004. "Redox-Active Metals, Oxidative Stress and Alzheimer's Disease Pathology," *Ann. N. Y. Acad. Sci.,* 1012, 153-163.

James, G. M., 1954. *Stolen Legacy.* New York: Philosophical Library, United Brothers Communication Systems.

Jenner, P., 1989. "Clues to the Mechanism Underlying Dopamine Cell Death in Parkinson's Disease," *J. Neural. Neurosurg. & Psychiatry,* 22, 28.

Jenner, P., and Olanow, C. W., 1996. "Oxidative Stress and the Pathogenesis of Parkinson's Disease," *Neurology,* 47 (suppl. 3), S161-S170.

Kastner, A., Hirsch, E. C., et al., 1992. "Is the Vulnerability of Neurons in the Substantia Nigra of Patients with Parkinson's Disease Related to Their Neuromelanin Content?" *Neurochem.*, 59, 1080-1089.

Kelawata, N. S., and Anathanarayan, L., 2002. "Antioxidant Activity of Selected Foodstuffs," *Internal. J Food Science and Nutrician,* 55, 511-516.

King, R., 2001. *Melanin: A Key to Freedom.* Chicago: Lushena Books.

King, R. D., 1990. *Selected References to the Eye of Herufrom Pyramid Texts.* Durham, NC.

--1994. *The African Origin of Biological Psychiatry.* Hampton, VA: U. B. and U.S. Communications.

Kozlowsky, G. P., Scott, D. E., and Dudley, G. K., 1973. "Scanning Electron Microscopy of the Third Ventricle in Sheep," *Z. Zellforsche,* 136, 169-176.

LaCerra, P., and Bingham, R., 1998. "The Adaptive Nature of the Human Neurocognitive Architecture. An Alternative Model," *Proc.Natl. Acad. Sci. U S. A.,* 95, 11290-11293.

Lacy, M. E., 1984. "Phonon Electron Coupling As a Possible Transducing Mechanism in Bioelectronic Processes Involving Neuromelanin,"*J Theor. Biol.,* 111, 201-204.

Lang, A. E., and Lozano, A.M., 1998. "Medical Progress: Parkinson's Disease," Parts 1 & 2, *New England Journal of Medicine,* 339,1130-1143, 1040-1053.

Larsson, B. S., 1993. "Interaction Between Chemicals and Melanin, *Pigment Cell Res.,* 6, 127-133.

Levi, A. C., DeMattei, M., et al., 1989. "Effects of 1-methyl-4-phenyl-1,2,3,6-tetrahydropyridine (MPTP) on Ultrastructure of Nigral Neuromelanin in *Macacafascicularis,*" *Neurosci. Lett.,* 96, 271-276.

Lim, G. P., Chu, T., Yang, F., Beech, W., Frautschy, S. A., and Cole, G. M., 2001. "The Curry Spice Curcumin Reduces Oxidative Damage and Amyloid Pathology in an Alzheimer's Transgenic Mouse," *J Neuroscience,* 21, 8370-8377.

Lin, J. K., et al., 1994. "Molecular Mechanism of Action of Curcumin," in *Food Phytochemicals II: Teas, Spices, and Herbs, American Chemical Society,* 20, 196-203.

Lindquist, N. G., 1987. "Neuromelanin and Its Possible Protective and Destructive Properties," *Pigment Cell Res.,* 1, 133-136.

Magavi, S. S., and Macklis, J. D., 2001. "Manipulation of Neural Precursors in Situ: Induction of Neurogenesis in the Neocortex of Adult Mice," *Neuropsychopharmacology,* 25, 816-835.

Mahady, G. B., Pendland, S. L., Yun, G., and Lu, Z. Z., 2002. "Turmeric *(Curcuma longa)* and Curcumin Inhibit the Growth of *Helicobacter pylori,* a Group 1 Carcinogen," *Anticancer Res.,* 22, 4179-4181.

Mann, D. M., and Yates, P. 0 ., 1983. "Possible Role of Neuromelanin in the Pathogenesis of Parkinson's Disease," *Mech. Age Dev.,* 21, 193-203.

Marsden, C. D., 1983. "Neuromelanin and Parkinson's Disease,"*J. Neural Transm.* (suppl. 19), 121-141.

Masliah, E., et al., 2000. "Dopaminergic Loss and Inclusion Body Formation in Alpha Synuclein Mice: Implications for Neuro degenerative Disorders," *Science,* 287, 1265-1269.

Mason, H. S., 1959. "Structure of Melanins," in Myron Gordon, ed., *Pigment Cell Biology.* New York: Academic Press.

McGinness, J., 1985. "A New View of Pigmented Neurons," *J. Theor. Biol.,* 115, 475-476.

McGinness, J., Corry, P., and Proctor, P., 1974. "Amorphous Semiconductor Switching in Melanins," *Science,* 183, 853-855.

Moore, T. 0., 1995. *The Science of Melanin. Dispelling the Myths.* Venture Books/Beecham House Pub.

-2002. *Dark Matters, Dark Secrets.* Redman, GA: Zaman Press.

Natarajan, C., and Bright, J., 2002. "Curcumin May Block Progression of Multiple Sclerosis." Annual Experimental Biology 2002 Conference, New Orleans, LA.

National Institute of Environmental Health Sciences [NIEHS], "Fact Sheet, Parkinson's Disease Research," April 1999.

Nicholaus, R. A., 1997. "Coloured Organic Semiconductors: Melanins," *Rend. Acad. Sci. Fis. Mat.,* 46, 325-340.

Nicholas, R.A., Patel, M., and Fattorusso, E., 1964. "The Structure of Melanins and Melanogenesis IV," *Tetrahedron,* 20, 1163.

Nicholas, RA., and Piattelli, M., 1965. "Progress in the Chemistry of Natural Black Pigments." *Rend Acad. Sci. Fis. Mat.,* 32, 83-97.

Noden, D., 1975. "An Analysis of the Migratory Behavior of AvianCephalic Neural Crest Cells," *Devel. Biol.,* 42, 106-130.

Odh, G., Carstam, R., et al., 1994. "Neuromelanin of the Human Substantia Nigra: A Mixed Type Melanin," *J Neurochem.,* 62, 2030-2036.

Ono, K., Hasegawa, K., Naiki, H., and Yamada, M., 2002. "Curcumin Has Potent Anti-Amyloidogenic Effects for Alzheimer's Beta Amyloid Fibrils *in vitro*," *J. Neuroscience Res.*, 75, 742-50.

Pavan, W. J., and Tilghman, S. M., 1994. "Piebald lethal (sl) Acts Early to Disrupt the Development of Neural Crest Derived Melanocytes," *Proc. Natl. Acad Sci. US. A.*, 91, 7159-7163.

Privat, A., and Leblond, C. P., 1972. "The Subependymal Layer and Neighboring Region in the Brain of Young Rats," *J Comp. Neurobiol.* 146, 277-301.

Robinson, K. P., 1979. "Electrical Currents through Full Grown and Maturing Xenopus Oocytes," *Proc. Natl. Acad Sci. U S. A.*, 76, 837-841.

Rugh, R., 1977. *A Guide to Vertebrate Development,* 7th ed., 56. New York: Macmillan.

Sapper, C.B., and Petito, C. K., 1982. "Correspondence of Melanin Pigmented Neurons in Human Brain with Al-Al 4Catecholamine Cell Groups," *Brain,* 105, 87-101.

Scapagnini, G., Colombrita, C., Calabrese, C., Pascal, A, Schwartzman,

M. L., and Abraham, N. G., 2004. "Curcumin Cytoprotective Effect in Rat Astrocytes ...," presented at Experimental Biology 2003 Conference, Washington, DC, April 17-21, 2004.

Schraermeyer, U., 1996. "The Intracellular Origin of the Melanosome in Pigment Cells: A Review of Ultrastructure Data," *Histol. Histopathol.*, 11, 445-462 (review).

Scott, D. E., Kozlowski, G. P., et al., 1973. "Scanning Electron Microscopy of Human Cerebral Ventricular System," *Z. Zellforschung,* 139, 64, 68.

Shu, S. Y., Wu, Y. M., Bao, X. M., et al., 2002. "A New Area in the Human Brain Associated with Learning and Memory: Immunohistochemical and Functional MRI Analysis," *Mal. Psychiatry* 7, 1018-1022.

Simonian, N. A., and Coyle, J. T., 1996. "Oxidative Stress in Neurodegenerative Diseases," *Annu. Rev. Pharmaco. l Toxicol.,* 36, 83-106.

Simon, H. H., Bhatt, L., Gherbassi, D., Sgado, P., and Alberi, L., 2003. "Midbrain Dopaminergic Neurons. Determination of Their Developmental Fate by Transcription Factors," *Ann. NY Acad. Sci.,* 991, 36-47.

Smith, U., 1970. "Aspects of the Fine Structure and Function of the Subcommissural Organ of the Embryonic Chick," *Tissue & Cell* 2, 19-32.

Spemann, H., 1938. *Embryonic Development and Induction,* Ch. VIII, 401. New Haven, CT: Yale University Press.

Sriram, K., Pai, K.S., Boyd, M. R., and Ravindranath, V., 1997. "Evidence for Generation of Oxidative Stress in Brain by MPTP: *in vitro* and *in vivo* Studies in Mice," *Brain Res.,* 749, 44-52.

Strzelecka, T., 1992. "A Band Model for Synthetic DOPA Melanin," *Physiol. Chem. Phys.,* 14, 219-233.

Sudkha, K., Rao, A., and Rao, S., 2003. "Free Radical Toxicity and Antioxidants in Parkinson's Disease," *Neurology India,* 51, 60-62.

Sulzer, D., Bogulavsky, J., etal., 2000. ''Neuromelanin Biosynthesis is Driven by Excess Cytosolic Catecholamines...,'' *Proc. Natl. Acad. Sci. US. A.,* 97, 11869-11874.

Swart, H. M., Sarna, T., and Zecca, L., 1992. "Modulation by Neuromelanin of the Availability and Reactivity of Metal Ions," *Ann. Neural.,* 32, S69, S75.

Synder, E. Y., Yoon, C, Flax, J. D., and Macklis, J. D., 1997. "Multipotent Neural Precursors Can Differentiate Toward Replacement of Neurons Undergoing Targeted Apoptosis Degeneration in Adult Mouse Cortex," *Proc. Natl. Acad. Sci. U S. A.,* 94, 11663-11668.

Tief, K., Schmidt, A., and Beermann, F., 1998. "New Evidence for Presence of Tyrosinase in Substantia Nigra, Forebrain and Midbrain," *Brain Res. Mal. Brain Res.,* 53, 307-310.

Weston, J.A., 1982. "Neural Crest Cell Development," *Prag. Clin. Biol. Res.,* 85, 359-379.

Yang, T., Lim, G. P., et al., 2004. "Curcumin Inhibits Formation of A-b-oligomers and Fibrils and Binds Plaques and Reduces Amyloid *in vitro,"* *J. Biol. Chem.,* 10, 1074 (online).

Yantiri, F., and Andersen, J. K., 1999. ''The Role of Iron in Parkinson's Disease and 1-methyl-4-phenyl-1,2,3,6-tetrahydropyridine Toxicity," *JBMB Life,* 48, 139-141.

Zecca, L., Macacci, S., Seraglia, R., and Parati, E., 1992. "The Chemical Characterization of Melanin Contained in the Substantia Nigra of the Human Brain," *Biochem. Biophys. Acta,* 1138, 6-10.

Zecca, L., Tampellini, D., Gerlach, M., et al., 2001. "Substantia Nigra Neuromelanin: Structure, Synthesis, and Molecular Behavior," *Mal. Pathol.,* 54, 414-418.

Zeevalk, G.D., Bernard, L. P., and Ehrhart, J., 2003. "Glutathione and Ascorbate. Their Role in Protein Glutathione Mixed Disulfide Formation during Oxidative Stress and Potential Relevance to Parkinson's Disease," *Ann. NY. Acad Sci.*, 991, 342-345.

CHAPTER 3

The Clinical Use of Bliss:
A Standardized Technique for Conscious Intervention Into the Functioning of the Autonomic Nervous System
Edward Bruce Bynum, Ph.D.

"All animals have an internal core of melanin in their brains. All humans possess this Black internal brain evidence of their common Black African origin. The all black neuromelanin nerve tract of the brain is profound proof that the human race is a Black race, with many variations of black, from Black-Black to White Black. One of the critical keys that distinguishes man from all other animals is this presence of intense blackness, neuromelanin pigmentation of the locus coeruleus, Black Dot, the upper most center of pigmentation, the doorway that opens into an all black hall of blackness, the neuromelanin "Amenta" nerve tract."

Richard D. King, M.D.
African Origin Of Biological Psychiatry

"Brain melanin (neuromelanin) increases with ascent up the phylogenetic ladder, reaching a peak concentration in man. Moreover, it is invariably found in strategic highly functional loci of the brain Neuromelanin within neurons and glia is concentrated in strategic locations of the brainstem which (together with monoaminergic axonal and dendrite extensions) allow for the "gating" of all sensory and motor input and output as well as all emotional and motivational input and output."

F.E. Barr, M.D.
Melanin: The Organizing Molecule

Address all correspondence to Edward Bruce Bynum, Ph.D., ABPP, Director of Behavioral Medicine, University of Massachusetts, Health Services, 127 Hills North, Amherst, MA 01003, USA

The author is indebted to the following people for their technical comments and expertise in the review of this chapter:

T. Owens Moore, Ph.D., Professor of Psychology, Clark Atlanta University and the Neuroscience Institute, Morehouse School of Medicine;

Alan J. Calhoun, M.D., Medical Director, University of Massachusetts, Health Services;

Warren H. Morgan, M.D., Medical Director, Amherst College Health Services;

A standardized clinical technique for conscious and directive intervention into autonomic nervous system functions is presented along with a description of both subjective and psychophysiological reactions to this technique. Its use with numerous symptoms and conditions is discussed as a means to decrease stress-producing somatic dysfunctions and increase focused relaxation of psychophysiological processes. Potential pathways and underlying mechanisms of the technique and operations are also discussed as they relate to the mid brain limbic system and the somatic unconscious. The process of physiognomic perception is discussed. The role of neuromelanin in somatic and other nervous system operations is elaborated.

KEY WORDS: Kemetic, neuromelanin, melanin, autonomic nervous system, limbic system, alternate nostril breathing, meditation, hemispheric dominance, physiognomic perception, acupuncture, nadis, psychophysiological equipoise, locus coeruleus, biological superconductivity, quantum resonance, Dogon of Mali.

Overview and Statement of the Problem

In many different clinical fields, a skilled intervention into autonomic nervous system functioning is crucial for therapeutic work. This is true whether working in the field of behavioral medicine, in biofeedback specifically, or in the related field of clinical hypnosis. In each one of these fields, the autonomic nervous system is often a focus for intervention since its delicate balancing operation is seen to underlie the manifestation of somatic symptomatology.

With so many diverse procedures and practitioners in these clinical fields, there is a need for a standardized technique for conscious and skilled intervention into the autonomic nervous system. This technique or procedure would need to be capable of selectively influencing the autonomic nervous system. In addition, the technique should be demonstrated to the patient with only minimal regard for experimental and experimenter bias. The technique should be teachable in a *standardized* way in order to elicit *standardized responses* from the patient. It is also preferable that it be a technique that can affect the autonomic nervous system while the patient is fully conscious and in control of psychological and physical faculties.

What follows is a brief chapter on a clinical technique that is in operation in specific therapeutic settings focused within a certain constellation of somatic symptoms. It will also offer the psycho physiological and psychological concomitants and repercussions of this technique. We will then discuss how this technique can be used in various clinical situations. Finally, this chapter will propose some ideas as to why this technique is clinically effective. We will also discuss the psychological and psychophysiological basis for the efficacy of the technique and its relationship to other disciplines in the alteration and transformation of consciousness.

Brief History

The history of therapy and healing has always been intimately associated with the clinical practitioner's capacity to affect the autonomic nervous system (ANS) and human consciousness. Any successful therapeutic methodology has to offer an explanation to the patient or sufferer as to the how and why of their predicament. The explanation needs to be consistent with the worldview of the time and era of the therapeutic setting. In each setting, it was and is the clinical and ethical task of the clinician to listen closely to the patient and communicate verbally and nonverbally that someone with training and education cares about them and that their symptoms are explainable and their condition controllable or at least amenable to therapeutic influence. This communication allows the clinician a conscious entrance into both the patient's unconscious and autonomic system and also into what we would term the patient's interpenetrating bio-informational energy field.

The clinical settings of today often use" scientific" modern metaphors such as computer analogies, conscious/unconscious dichotomies, and other "modern language" to describe what is actually an ancient discovery (Bandier and Grinder, 1982; Bandier and Grinder, 1981; Hourning, 1986). Even in ancient Kemetic Egyptian times and then in later Greek and Roman times, the intervention into the autonomic nervous system, using various procedures such as clinical hypnosis, auto-suggestion, and other techniques to decrease or eliminate the numerous stress-related "diseases" and pain syndromes, was well testified to in the classical literature (Muses, 1972; Edmonston, 1986; Ebbell, 1937).

Due in large measure to the practice of mummification, the original indigenous Kemetic Egyptian knowledge of anatomy and physiology was quite extensive and precise, only to be surpassed three millennia later in the 18th century in Europe (Finch 1990). Much as their practices laid the foundation and template of human civilization (Diop, 1974; Jackson, 1970), their developing practices and knowledge of the body essentially became the template of Western scientific medicine. With the advance of modern anatomy, neuroanatomy, and psychophysiology, however, we have been able to progressively and more clearly delineate exactly the psychophysiological processes that are involved in these procedures.

Clearly, many of the ancient cultures had sophisticated methods for the treatment of a wide variety of symptoms, symptoms we still see in clinical practice today (Finch, 1990). The technique we will elaborate here draws from many different sources, but in particular modern psychophysiology, neuroanatomy, certain Yoga practices, an awareness of specific acupuncture correspondences, and some forms of a higher cortical skill or "emergent mental disciplines" known as meditation. In concert, these disciplines can exercise a controlling influence on lower-order symptom phenomena operating in the body and mind. Given, as earlier chapters have pointed out, the intimate interplay among melanin, neuromelanin, bioelectrical conductivity and the central nervous system, the connection between consciousness and neuromelanin will become more obvious as this chapter unfolds.

Description of the Technique

The technique itself falls into three phases. We will elaborate the psychophysiological basis of each of those phases as this article progresses. At the present time it should be noted that the three phases are comprised of: (1) diaphragmatic breathing; (2) alternate nostril breathing; and (3) a particular focus of attention on a specific area of the body followed by the "re-awakening" of this sensation in clinically focused regions of the body. These phases, when done in sequence, provide a replicatable demonstration of specific stimulation of the autonomic nervous system under the direct control of the person's attention and focus.

The first phase involves diaphragmatic breathing. This means directly teaching the patient by demonstration on oneself to differentiate between chest, or thoracic, breathing and diaphragmatic breathing. In the process of doing this, the patient will experience an observable and undeniable sensation. The clinician can then draw the patient's attention to certain psychophysiological processes that are stimulated by diaphragmatic breathing. After the diaphragmatic breathing is sustained for 5 to 8 minutes, the patient is taught alternate nostril breathing.

Alternate nostril breathing, a well-established yogic technique that affects the ANS, serves to deepen and intensify the psychophysiological reactivity that has already been initiated by the diaphragmatic breathing (Kuvalayanandal,1978; Funderbunk, 1977; Thakkur, 1977). The alternate nostril breathing is usually done at a slow pace for anywhere from 3 to 5 minutes. After alternate nostril breathing, which followed the diaphragmatic breathing, the patient is taught to relax and focus attention at a place on the top of the lip and bottom of the nose. The patient's attention is drawn subtly to the alternating current of slightly warm air, then cool, then warm again, which they can notice in the top of the lip, bottom of the nose area.

When this three-part pattern is done in sequence, the patient will exhibit a definite, observable psychophysical response that the clinician can then draw attention to the patient. It leads to a decidedly "blissful" feeling and sometimes to the actual organismic perception of a "blissful current" coursing through the body along a certain pathway. This blissful sensation in its clearest form is an on- ideational, somewhat luminous sensation not identified with any specific organ or site but one that may, when directed, pervade a particular region. It has an affinity to the sensation that occurs immediately anterior to a strong sexual orgasm and also to the somatic memory of the body and mind in deep sleep without dreaming, where there is little or no thought but certainly an intuition of bodily and psychological restoration and rejuvenation. We will have more to say about this later in this paper. This whole procedure takes approximately 10 to 12 minutes to perform successfully. After that point, the patient's attention is drawn clinically to the areas of the body-mind that are the focus for clinical intervention.

What follows is a two-part description of the psycho physiological and psychological experience of a patient executing this procedure. First, the subjective experience is presented.

Subjective and Somatic Experience of the Procedure

The procedure initiates a number of clearly observable subjective experiences on the part of the patient. The patient initially feels more "relaxed." In some patients, there is an initial paradoxical response of increased anxiety before relaxation. However, in the majority of patients relaxation is the first noticeable sign. With continual practice, the patient begins to notice that there is a gradual slowing down in the subjective perception of the "speed" of their thoughts. Also, the patient often spontaneously begins to notice more imagery emerge from their experience. Sometimes this imagery is associated with immediate experiences, and sometimes long forgotten memories come to the surface. The unconscious is accessed by this procedure.

In terms of their psychophysiological experience, a number of things become noticeable to the patient during the procedure. A large percentage of patients witness an increase in saliva. There is also an increase in heat on the surface of the skin due to vascular dilatation and an increase in GI tract activity in a pleasant direction, as though one were getting ready for a meal. The patient may also spontaneously notice that certain areas of the body, particularly the clinically effected areas, become clearer in terms of their visualization and organismic perception. Finally, a significant number of patients will notice a "tingling sensation" over the entire surface of the body, in particular around the face and limbs. These phenomena occur in varying degrees.

The clinician, while observing the patient, can identify some of these processes and selectively amplify or reinforce them for clinical effect. With this increase in psychophysical reactivity, shifts in somatic perception, and subjective experiences of relaxation and even euphoria and blissful sensations at times, the patient can begin to notice the capacity for potentially altering sensory experience (Erickson and Rossi, 1979). Patients become engaged in this immediate experience and also, in conjunction with biofeedback equipment, can gain direct biomedical information concerning their symptoms and psychophysical reactions by way of proprioceptive feedback from certain areas of the body.

For instance, a patient may notice an increase in heat in localized regions of the body or generalized over the entire surface of the body. This is due to increased blood flow or vascular dilatation. This is useful when certain symptoms involving circulatory or cardiovascular problems are the target of clinical focus, e.g., common and classical migraine headaches, labile high blood pressure, dysmemorrhea, Raynaud's disease, Berger's Disease, and other related phenomena. There is also a decided increase in the feeling of identification with and subjective relaxation in certain areas of the body. This is due to the relaxation of specific muscle groups. This is a useful technique in dealing with issues involving muscular and motoric constriction and/or spasticity, e.g., GI tract disturbances (IBS, etc.), muscle contraction headache, chronic pain syndromes, etc.

However, the procedure is not limited to these systems. Others, such as the respiratory system, are also affected, making it possible to treat asthma, rhinitis, and related ailments. A number of clinical studies have indicated that an increase in relaxation and a decrease in psychophysical reactivity and stress chemistry lead to an increase in immune-enhancement or immune functions (Locke and Colligan, 1986; Kiecolt-Glaser, et al., 1985; Achterberg and Lawlis, 1984; Rider, et al., 1990). It is well known that harmful stress hormones are decreased by calm imagery and thought and that pain-killing endorphins and the immune system can be modulated by therapeutically guided mental states. There is enormous potential for the patient. This technique allows the patient, with progressively deepening skill and "emergent" higher-order cortical control processes, to interact consciously and specifically with the soma and thereby literally "talk" to certain areas of the body. By teaching the patient a skillful observational method and a way to intervene in the psycho-physiological process of the body, the patient learns to selectively enhance or decrease the reactivity of certain somatic functions. In this process, the patient spontaneously remembers or associates in an affectively charged way to certain images, ideas, and motifs that can be clinically quite useful in a psychotherapeutic context. The psychological and somatic regions of the unconscious are accessed. There is a certain affinity here between clinical hypnosis, imagery, and auto- suggestion. In the context of biofeedback, however, these can be enhanced significantly and accurately due to the nature of the psycho-physiological monitoring feedback process.

It is very important to note at this juncture that the surface of all these internal organs, along with the entire surface of the brain, crucial regions of the mid-brain, parts of the endocrine system, autonomic nervous system, and peripheral nervous system, contain significant amounts of semi-conducting melanin and neuromelanin as part of their structure and content. It is suggested here that in ways we are only now beginning to understand this complex and interwoven bioinfomlationa l field, generated in early embryogenesis and developing throughout the life cycle, communicates with itself via conductivity and sometimes, under the right conditions, consciousness.

Psychophysiological Description of the Procedure

What follows is a closer and clearer description of each one of the three sequential stages of the technique. Each one will be presented in more detail along with its psycho-physiological basis.

The first phase of the procedure is referred to as diaphragmatic breathing. Diaphragmatic breathing, as opposed to chest, or thoracic breathing, involves the movement of the abdomen and the diaphragm in the breathing process (Fried, 1987). Breathing air into the deep recesses of the lungs is almost always a healthy activity. The pericardium is attached to the diaphragm, and thus the process of deep breathing causes the diaphragm to descend, stretching the heart slightly downward toward the abdomen. When the lungs are filled with air from the bottom upward, they compress the blood-rich viscera, giving a gentle massage to the heart and the internal organs. As the diaphragm contracts and releases, it also massages the heart, pancreas, liver, stomach, small intestine, abdomen, and other internal organs. This leads to a better diffusion of blood through the system and a gentle stimulation of the internal organs (Lysebeth, 1983; Rama, Ballentine, and Hymes, 1979). This internal process affects the autonomic nervous system with observable results.

The human nervous system can be divided structurally into the central and peripheral branches. The peripheral branch is subdivided functionally into the somatic branch, which is generally conscious, and the involuntary, or autonomic, branch, which is generally unconscious. The autonomic nervous system itself is subdivided into two branches, the sympathetic and the parasympathetic nervous system branches. See Chart I.

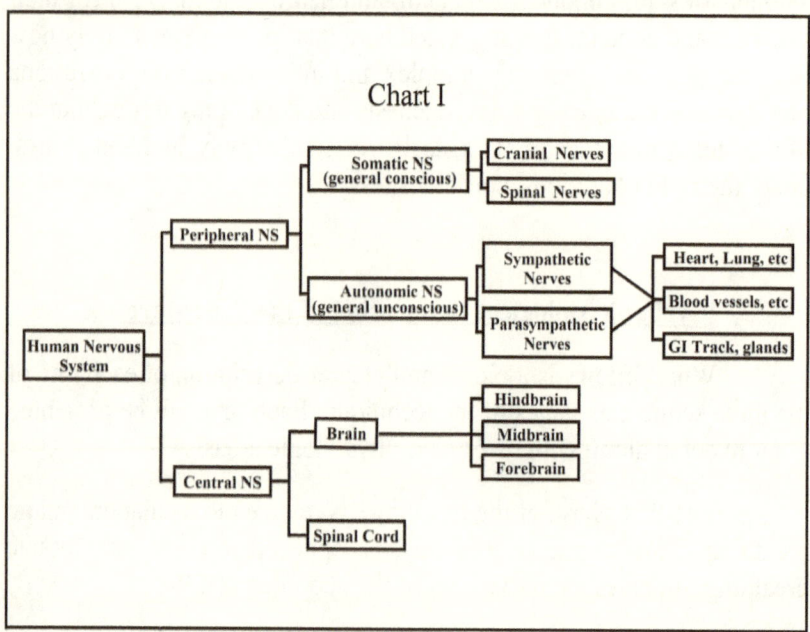

Chart I

These branches usually work in balanced opposition with each other so that an overall harmonious regulation results. For example, in cardiac functions the parasympathetic system focuses on slowing down the heart rate while the sympathetic system is involved in increasing the speed of the heart rate. Their dynamic balance is the ideal for lowered levels of stress. Symptoms often arise when there is a sustained imbalance in either one. In the case of accelerated heart rate (tachycardia) or slowed heart rate (bradycardia), arrhythmias of several kinds may eventually arise resulting in a variety of well-known clinical syndromes. Psychological stress and/or depression can lead toward certain arrhythmias, innumerable psychosomatic disorders, or the suppression of immunocompetence systems (Locke and Colligan, 1986; Achterberg and Lawlis, 1984). See Chart II.

The sympathetic nervous system consists mainly of two vertical rows of ganglia, or constellations of nerve cells, arranged on either side of the spinal column. Their branches spread out to all of the different organs, glands, and internal systems of the abdomen, thorax, and other areas of the body. They also intermingle in integrated plexuses with nerve branches of the parasympathetic system. It is significant that a principal part of this system is the tenth cranial nerve, also called the vagus, or wandering nerve. It is connected to the hind-brain and travels downward along the spinal cord through the neck, chest, abdomen, and other vital organs, sending out its branches into various nerve constellations with the sympathetic system. The vagus ends in a constellation, which is connected to the solar plexus. However, even though it may end in the solar plexus, it still sends thin filaments to lower levels of the body (Netter, 1972; Tokay, 1972).

In clinical practice, when the patient begins to slowly and methodically regulate the motion of the lungs, the heart itself becomes slowly regulated. Eventually, the right vagus nerve is brought under conscious control and thereby the area of the brain that is implicated in the involuntary or autonomic systems of the body is made amenable to conscious influence by the patient (Rama, Ballentine, and Hymes, 1979).

To the extent that the mind becomes focused, it then is capable of extending or amplifying its capacity to volitionally influence the body.

The metaphor of a laser that focuses the diffuse light of the mind or consciousness into a coherent light that has different capabilities is often used with the client to demonstrate this principle. From this vantage point, there is no area of the body-mind that is not in some degree amenable to conscious influence by the higher cortical centers of the brain and consciousness. Eventually in the process, the patient's attention is drawn more and more into the exhalation phase of the breathing cycle. This gentle emphasis on exhalation differentially reinforces the parasympathetic branch of the autonomic nervous system, leading to a decrease in conventional stress for most of the organ systems (Rama, Ballentine, and Hymes, 1979).

The autonomic nervous system has a complicated series of interactions and innervations with the different organ systems of the body. As a general rule, the sympathetic system creates tension and constriction in the systems, and the parasympathetic system creates relaxation or dilatation. There are a few exceptions to this generalization. The pupils of the eye and the bronchi of the lungs initiated by intense interest or anxiety are examples of the opposite of these tendencies. See Chart 1.

After the patient has mastered diaphragmatic breathing his/ her attention is next drawn to alternate-nostril breathing. It is of interest that the nasal passages have alternate cycles of dominance of approximately 11/2 to 13/4 hours (Klein and Armitage, 1979). In other words, for approximately an hour and a half to an hour and forty-five minutes either the left or right nostril is more dominant, allowing air to pass more freely through its turbinates. The nasal system is actually a neurological system, in that the first cranial nerve, the olfactory used for smelling, has subtle nerve endings in the top of the mouth, bottom of the nose area. Originating in the nasal mucosa, these olfactory cells are presently the only known sensory cells to conduct impulses. They course through the cribriform plate to the olfactory bulb, then backward and along the olfactory nerve below the frontal lobe, dividing into two branches. It is unclear whether the medial branch in humans ends in the subcallosal gyrus or the far olfactory area. The lateral branch does terminate in the uncus and, very importantly, in the emotional and "meaningful" memory- processing hippocampus gyrus (Netter, 1972).

It is curious and clinically significant that each time we inhale and exhale we stimulate this nerve, which sends a subtle message to the mid-brain limbic system, the system that supervenes over our primitive emotionality, e.g., flight-fight response, disgust, gustatory, etc. Furthermore, this mid-brain limbic system, consisting of the amygdala, hippocampus, and hypothalamus, is deeply implicated in the generation and filtering of powerful unconscious emotions ranging from lust to rage to murderous impulses to spiritual intoxication and anomalous experiences that alter our perception of space, time, location, and reality (Joseph, 2000). Many primitive emotions and reactions and their coordinated bodily functions, from respiration to cardiac functions to gastrointestinal motility, are stimulated in this way. That is why we look to the limbic structures for the root of this bodily current of primal feeling. It is essentially our "emotional body".

When the tenth cranial nerve, or vagus nerve in particular, is stimulated by way of this deep, diaphragmatic breathing with an emphasis on the exhalation phase of the breathing cycle, the hormone epinephrine, which arises in the body outside these neural structures, and the neuromodulator norepinephrine, which arises within these neural structures, are brought into interaction (Hassert, Miyashita, and Williams, 2004). This release of norepinephrine quickly floods the amygdala, thereby stimulating and deepening the memory of powerful positive or negative emotions and experiences from the "emotional body".

Because the hormone epinephrine cannot cross the blood brain barrier that enfolds these mid-brain structures, the ascending or afferent fibers of the vagus nerve that pick up the stress chemistry produced by the adrenal medulla in the flight- fight response stimulate the neurons in the brainstem, known as the nucleus of the solitary tract (Hassert, Miyashita, and Willams, 2004). From there, the message is transmitted to the ancient mid- brain limbic system for "meaningful" consolidation by both the amygdala and perhaps the hippocampus.

These mid-brain limbic structures that communicate feeling and emotion then can flow, interface with bodily sensations, and "project" to other areas of the brain. The motor projections from the spinal lineup are to the anterior or pre-central gyrus, and the sensory projections are to the post-central gyrus. Both gyri have the body topologically outlined on the surface. There are innumerable connections by nerve fibers of varying degrees of my elimination and by long, pyramidal tracks that twist and turn through complex and curving convolutional "spaces" before they reach their final destination in the projection areas of the cerebral cortex. It is truly awesome to see how these interweaving, nonlinear spaces and fibers have woven together the different regions, spaces, and functions of the cerebral universe.

Powerful emotions and feelings have been deeply rooted in hominid neurobiological structures for millions of years. Archaic humans had religious impulses, buried their dead, and experienced the full spectrum from awe to terror. Before the emergence of complex thought, speech, and linguistic patterns, they had created tools and survival skills, made strong emotional attachments, and lived through lust, passion, and dreams. These are rooted in the temporal lobes and mid-brain limbic system. Only later, as Homo sapiens evolved toward Homo sapiens sapiens, our species, did there develop the capacity for initiative, long-term goal setting, and cognitive reflection on multiple options in any given situation. The size of the frontal lobes has increased by 33 percent in Homo sapiens sapiens, while the limbic structures have remained essentially unchanged (Joseph, 2000). The deep-core emotions of embodiment are an evolutionary, neurobiological unfoldment of these limbic structures. What we outline here is a conscious way to intervene into this system.

By careful diaphragmatic breathing and alternate nostril breathing, the patient sends, by way of the first cranial nerve and the tenth cranial nerves of the somatic nervous system, a specific and soothing message to the autonomic nervous system. The net result is continual vascular dilatation, increased heat on the surface of the body, continual blood diffusion through the body, more efficient use of the lungs, and massage of the internal organs. The net effect is a more

positive, consistent, and "embracing" proprioceptive feedback to the body. This significantly decreases the stress reaction and the harmful effects of constant stress chemistry to one's psychophysiology. As dreams were the "royal road" to the unconscious in early psychoanalysis, conscious and disciplined respiration is the velvet path into this "somatic unconscious." Both reflect and are implicated in powerful emotional processing that occurs in the limbic and paralimbic systems of the brain.

Many experimenters have come to notice that the nasal cycle opens a certain "window" on the autonomic nervous system and allows an opportunity for cerebral hemispheric influence to occur. Researchers have noticed that there is a direct relationship of cerebral hemispheric activity, as monitored by an electroencephalogram (EEG), and the ultradian rhythm of the nasal cycle (Werntz, 1981). Essentially, the relatively greater integrated EEG values in one hemisphere are positively correlated with the predominate airflow in the contra-lateral, or opposite, nostril (Werntz, 1981; Werntz, et al., 1981).

Researchers have also noticed that by changing nasal dominance by the technique of forced single-nostril breathing through the nondominant closed nostril, effects appear on the EEG. Thereby *one can experimentally shift the nasal dominance, and this is accompanied by a shift in cerebral hemisphere dominance to the contra-lateral hemisphere.* This has been known clinically for the last decade. However, it has also been the testimonial of certain meditative disciplines, in particular the Swara yogis (Prakashan, 1980), for thousands of years (Werntz, et al., 1981). This phenomenon, clinically speaking, eventually allows the patient to voluntarily change the relative focus of activity in the higher cortical centers of the brain and thereby influence the all-pervasive and supervenient autonomic nervous system reactions that regulate practically every major function of the body. By the technique of alternate nostril breathing, the patient is able to differentially affect the right or left hemisphere of the brain and its associated cognitive activities.

There is some controversy as to the psychophysical basis for this reactivity. However, most researchers believe that this technique represents an extensive integration of autonomic and cerebral cortical activity (Klein and Armitage, 1979). It is held that the nasal cycle is itself regulated centrally by the hypothalamus, thus altering the sympathetic/parasympathetic balance. This reaction occurs throughout the body, including the brain, and is perhaps the mechanism by which all basal motor tone regulates the control of blood flow through the cerebral vessels, thereby itself altering cerebral hemispheric activity. This clearly is an influence, consciously, of hemispheric laterality and also of the limbic system upon the body by the higher cortical centers of the brain. The revolution in cognitive psychology has now established that subtle higher cortical processes can act as causal constructs in brain behavior and that, indeed, these mental states and skills as emergent properties of neurological activity, definitely exercise a supervenient influence over lower-order biological and psychological events (Sperry, 1988).

Regulating the system, quieting external noise or distractions, and intensifying internal concentration or attention accomplish the specific influence that is exercised by this procedure. Attention to external events and stimuli is simultaneously decreased. With this done systematically, various areas of the system are brought consciously under more and more influence and control of consciousness. Thereby, symptoms can be affected quite specifically.

The third phase of the procedure involves focusing attention on the top of the lip and the bottom of the nose. Focusing here is done usually on the exhalation phase of the breathing cycle. The patients are often asked to lick the top of their lips with their tongue, creating a little moisture so that the alternating current of warm, then cool, then warm air again, can be noticed. After that, progressively more attention is drawn toward the exhalation phase of the breathing cycle. The top of the lip, bottom of the nose area is chosen for two distinct reasons. The more overt reason is that the air current can be directly experienced here with relatively little effort. The second reason is that it tends to steady attention. But there is another reason for focusing here.

In acupuncture theory there is a major meridian that terminates at the area at the top of the lip, bottom of the nose. It is referred to as the "governor vessel meridian" (Motoyama, 1981). By focusing attention in this area, this subtle acupuncture meridian is technically activated and what is called the "chi," or "ki," energy is then moved more systematically, like a river, through the 12 ordinary meridians of the system. The principle underlies much of well-tested and centuries old acupuncture theory and practice (Huang Ti, 1966). This procedure also tends to steady attention and allow a pleasurable, even "blissful current" of non-ideational feeling and sensation to arise and organismically be perceived to move along the body axis of the spine. It tends to decrease any dominance of the sympathetic or the parasympathetic system and brings them more into a dynamic balance.

Finally, it should be noted that in some yogic meditative disciplines this procedure, which follows alternate nostril breathing, helps "depotentiate" the dominant tendency of the right or the left side of the body, and thus to allow for an opening of what is termed the "central canal" (Motoyama, 1981; Rama, 1981). There is the belief in yoga practice that the right and left sides of the body along the spine have a conduit called the Ida and Pingala Nadis for the movement of energy. They correspond, it seems, to the second lines of the urinary bladder meridian of acupuncture. When depotentiated, the current moves into this central canal, the body axis along which a pleasant, blissful current of sensation is organismically perceived to move. The dominance of either the "right side" sympathetic nervous system innervations or the "left side" parasympathetic system innervations is not conducive to deeper meditative states.

When the three procedures are sequentially followed and the final procedure is accomplished, it is often useful to have the patient begin to focus their sense of internal relaxation and psychophysical equipoise on the affected organ. In this undertaking, the eyes are slightly open and internally focused on the area or closed but still focused on the area and an inner part of the body. This procedure is very similar to the meditative procedure of Shambhavi mudra (Svatmarama, 1971). Subjectively, this is a very powerful procedure and clinically will differentially activate or "awaken" certain areas of the body, both subtle and gross.

It is important to note that some patients occasionally slip into the deeper regions of this practice and experience phenomena encountered by practitioners of other associated cognitive and contemplative disciplines. In particular, when the heart rate is slowed enough and coordinated with respiration, the heart-aorta system tends to produce an oscillation of about 7 hertz (Hz) that then reverberates throughout the entire skeletal framework, especially in the dense structures of the skull. This in turn is believed to create a series of standing waves projected to the ventricles of the brain, stimulating the perception of a moving sensory current that flows through the entire body but which is actually focused in the cortical and sub-cortical structures. These standing waves are in multiples of the 7 Hz base heart-aorta oscillation (Sannella, 1987; Bentov, 1977).

Other wider environmental entrainment effects have also been observed. This experience can be both intensely pleasurable but also disorienting and so should be avoided unless the patient is also practicing a specific contemplative discipline that is aware of these phenomena. The boundary between phenomena that are internal and external to the patient or practitioner also tends to become more blurred and porous, and thereby the perception of inner life, activity, and consciousness is easily perceived or recognized as existing throughout one's external space. "Space" itself tends to undergo various alterations at times, an interesting area that we will return to later in this chapter. The successful execution of this procedure in a primarily clinical context, however, gradually brings more of the unconscious areas of the body into conscious availability and thus susceptible to amelioration by biofeedback and behavioral medicine procedures. A number of clinical problems thereby become amenable to treatment. These include not only the symptoms associated with. cardiovascular activities, e.g., angina pectoris, arrhythmias, migraines, elevated blood pressure, etc., but also symptoms associated with muscular activities, e.g., TMJ, bruxism, muscle contraction headache, ulcerative colitis, IBS, etc., and potentially those of "soft" neurological and immunological significance (see Chart II).

Chart II. Partial List of Stress Modulated Diseases and Symptoms

I.   Cardiovascular Diseases
A.   Cardiac Arrhythmias
B.   Cerebral Stroke
C.   Angina Pectoris
D.   Coronary Artery Disease
E.   Hypertension
F.   Raynaud's Disease
G.   Migraine Headaches

II. Muscle-Related Symptoms
A.   Tension Headaches
B.   Oral Conditions.
1.   Bruxism
2.   TMJ
3.   Clenching
4.   Myofacial Pain Syndrome
5.   Necrotizing Ulcerative Gingivitis
6.   Apthous and Herpetic Lesions
C.   Shoulder Aches (Chronic)
D.   Backaches
E.   Neck Aches

III. Gastrointestinal
A. Colitis
1. Ulcerative (Inflammatory)
2. Spastic or Mucous
B. Peptic Ulcer
C. Fecal Incontinence
IV. Genitourinary
A. Impotence
1. Primary
2. Secondary
B. Dysmenorrhea and Amenorrhea

C. Dyspareunia and Vaginismus
D. Eneuresis and Encopresis
V. Allergic Diseases
A. Asthma
B. Chronic Urticaria (Hives)
C. Angioneurotic Edema (Allergic Swelling)
D. Vasomotor Rhinitis

VI. Infectious Diseases (Through stimulation of the immune system)

VII. Hyperthyroidism
VIII. Rheumatoid Arthritis

IX. Diabetes Mellitus (Through disordered metabolism and hyperglycemic disregulation).

X. Cancer(s) (By way of immune system compromises)
XI. Psychological-Psychiatric
A. Depression
B. Insomnia(s)
C. Anxiety Reactions
D. Behavioral Dysfunctions, Tics, etc.
E. Phobias

XII. Miscellaneous Diseases
A. Neurodermatitis (Skin Rash)

The Clinical Use of Bliss:

B. Alopecia (Hair Loss)
C. Graying hair (Prematurely)
D. Hypoglycemia
E. Thrombophlebitis

Discussion and Implications: The Somatic Unconscious

What follow are some preliminary ideas on the potential neuroanatomical pathways and psychophysiological activity that may accompany the aforementioned technique. We have already mentioned the psychophysiological reactions that occur when one is practicing this procedure. Psychophysiological activity is associated with the sequence of diaphragmatic breathing, alternate nostril breathing, and the focus of attention at the top of lip, bottom of nose area. It is suggested that the emergent higher cortical functions of the human nervous system in this context exert a decided influence on lower-order cognitive and psychophysical activity (Sperry, 1988). See Graph 1, a sagittal section of the human brain from cerebral cortex to cerebellum, including limbic system structures.

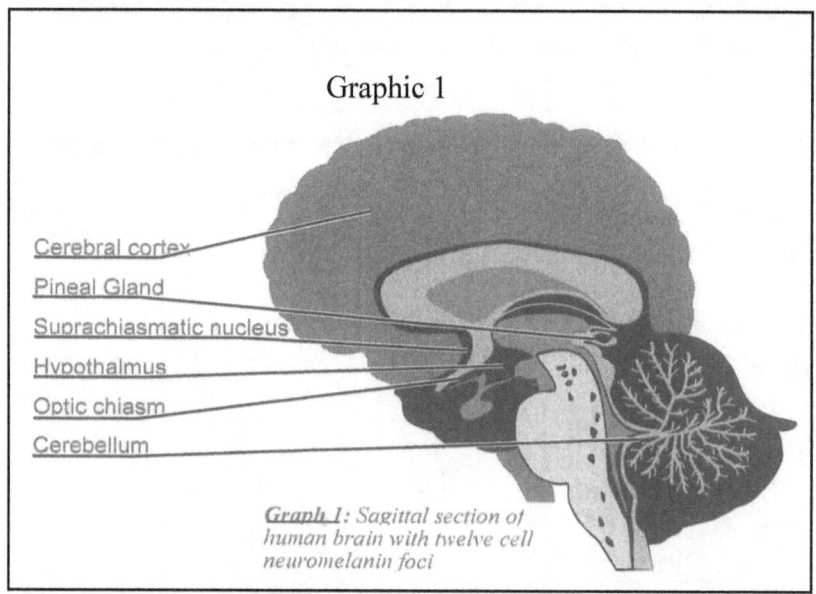

Graph 1: Sagittal section of human brain with twelve cell neuromelanin foci

Graph: *Sagittal Human Brain with neuromelanin*

In particular, the human nervous system, harnessing the processes of physical "tension" and psychological "attention," is able to differentially affect the peripheral and the central nervous systems. In reference to the peripheral nervous system, it is generally divided into the somatic and the autonomic nervous system (see Chart I). The somatic nervous system itself gives rise to the 12 cranial nerves and the 31 pairs of spinal nerves. The autonomic nervous system, however, a system that is generally unconscious, is involved with the sympathetic and the parasympathetic nervous system.

The sympathetic and parasympathetic nervous systems innervate all of the central organs of the body, including the heart, lungs, blood vessels, GI tract, etc. By learning how to differentially increase or decrease stimulation to these areas, using imagery and the control of respiration and attention, the first cranial, or olfactory, nerve to the mid-brain limbic system and the tenth cranial, or vagus, nerve is stimulated, and thereby a certain psychophysical reactivity is affected.

In terms of psychiatric symptomatology, it is notable that a specific cerebral area of the central nervous system, in particular the locus coeruleus, seems to account for approximately 70 percent of all central nervous system noradrenergic activity. King (1990) and others have pointed out that neuroanatomically, the locus coeruleus occupies the uppermost point in an all black neuromelanin nerve track that runs from the brain stem into the spinal cord. It has been observed that higher levels of locus coeruleus activity are correlated with hyper vigilance and attention to unusual or fear-provoking stimuli, but less activation appears to be associated with behavior such as sleep, grooming, and feeding (Foote, Bloom, and Aston-Jones, 1983). This would suggest that the locus coeruleus, an area highly associated with neuromelanin activity in the brain, acts as a controlling gate to stimuli coming into the system. In other words, the signal-to-noise ratio of incoming stimuli is significantly affected by the operation of the locus coeruleus. It appears to have a rather substantial afferent input from the internal organs and seems to receive direct innervations from the medullar nucleus solitarus (Elama, Svensson, and Thoren, 1986; Charney, Heninger, and Breier, 1984).

The medullar nucleus solitarus is an area known to be a principal locus for afferent information from the internal organs. Emotional modulation of this area due to stress, familial dynamics, or other stimuli and the perception of this area by the conscious mind produce the emotions of anxiety, fear, and potentially depression. Anxiety, depression, and fear can lead to hypo-motility while aggressive feelings such as hostility and resentment can cause hyper-motility(Gillis, Quest, Pagani, et al., 1991).

Psychological disorders that involve anxiety and somatization may be reflected in gastrointestinal and chronic pain syndromes. Studies have suggested that there is a pathological disregulation of the locus coeruleus in panic disorder (Gorman, Liebowitz, Fyer, et al., 1989) and possibly in depression (Charney, Heninger, and Breier, 1984; Charney and Heninger, 1986; Siever, Uhde, Jimerson, et al., 1984).

Essentially then, the locus coeruleus, an area associated with high levels of neuromelanin activity, is a possible central nervous system area having both afferent and efferent connections to the internal organs, and it could provide a pathway for some of the neuroactivity of the system.

The locus coeruleus may combine not only internal or visceral stimuli but also mid-brain limbic and higher cortical stimuli and then integrate and redistribute the information to other systems, all usually occurring outside of conscious awareness. These other mid-brain limbic system expressions are what is termed the somatic unconscious. See Graph 2 of primary limbic system structures.

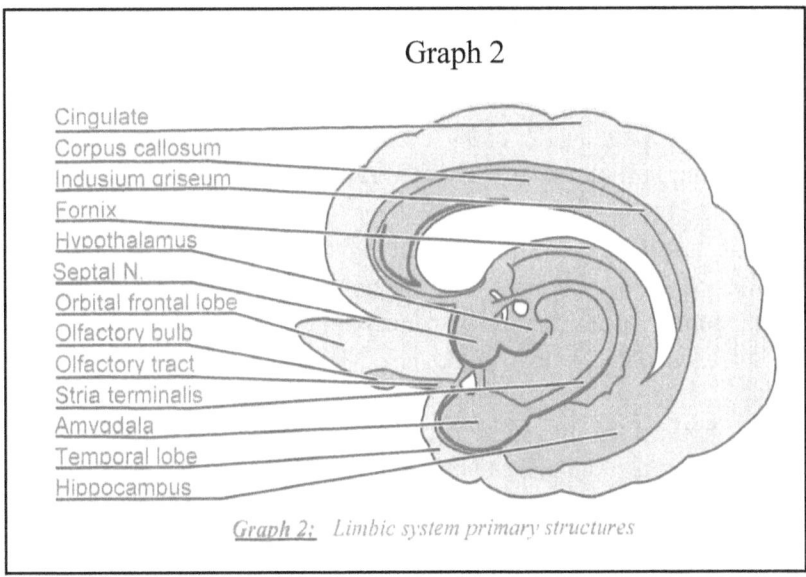

Graph 2: Limbic system primary structures

The locus coeruleus is a term of ancient lineage, deriving from the Latin word" locus," meaning point or dot, and the Sanskrit" caeruleus yamas," meaning black (King, 1900). This "black dot" area of the brain holds a significant amount of melanin (Amaral and Sinnamon, 1977). Its cells project to and provide for the primary noradrenergic nerve supply to many other sites in the brain, including the forebrain and the cerebral cortex, and also down to the hippocampus, the cingulate gyrus, and the amygdala areas, which comprise the major portion of the limbic cortex. This locus coeruleus is also known to supply part of the norepinephrine located in other brain sites, especially the hypothalamus, the thalamus, the habenula, or deep pineal, the cerebellum-lower brainstem, and also the spinal cord.

We mentioned above that neuromelanin appears to play a modulating role in this system (Moore, 1995, 2002). It is already known that neuromelanin in all likelihood plays a large role in the modulation of catecholaminergic transmission (Kagan and Rosenberg, 1987). The genetic, or biological, hypothesis presents itself here since neuromelanin is an integral aspect of the very earliest phases of human embryological development, beginning with precursors of the neural crest system itself, and it appears to exercise an organizing influence in this process (Barr, 1983). It is by no means conclusive, but it is certainly suggestive.

Actually, the deep brain core runs down and out of the brain stem nearly to the base of the spine. The spinal cord itself is an extension of that original embryonic neural crest. The spine is a longish white cylinder, oval in cross section. The inner matter is dark gray and the outer surface is white. In the brain itself this situation is reversed, and the outer surface is gray while the inner bulk is white. Modern neuroscientists have shown that the light, interacting melanin that creates the gray color of the brain is present in the brains of all animals, with the degree of pigmentation clearly increasing as creatures move up the evolutionary path (Marsden, 1961; Scherer, 1939). Mammals have the greatest pigmentation density; and primates have the highest among the mammals (Bogerts, 1981). Finally, even among the primates the higher the evolutionary form of brain complexity and organization, or similarity to the human type, the richer the light, interacting melanin and biochemistry of the brain.

We also know that melanin itself is highly concentrated not only on the surface of the brain and crucially in specific regions within the brain core, but also, clinically speaking, on the surface of many internal organs, the diffuse neuroendocrine system, and the sensory systems of the eye and ear. All of which communicate with each other via interconnected organismic pathways. This suggests a certain kind of energy and bio-informational field capable of self-regulation and conscious influence under certain conditions. This interpenetrating field responds to various forms of stimuli, in particular the potentially guiding stimuli of light, heat, sound, movement, and resonance. After all, melanin does absorb a rather wide spectrum of electromagnetic radiation and other energies of excitation from adjacent molecular structures. All this, mind you, occurs while shielded from the ultraviolet radiation of the sun by almost half an inch of skin and skull. Clearly, the brain, through the medium of melanin, absorbs radiation in its local aspects and perhaps even partakes in its nonlocal dynamics.

Consciousness, it seems, enters or projects into this three dimensional picture through its luminous, vibratory affinity with the dynamics of neuromelanin. In other words, at the level of neurodynamics, in behavior not of the cells but at the synaptic junctures, quantum mechanical processes seem to occur. These processes reflect the so-called "state vector" collapse of nonlocal light from its all pervasive vibratory "source" into localized actuality from the vast realm of many potential states existing in multiple or "N"dimensional space. This collapse of the nonlocal may even have an affinity for the localization of forms and "shapes" in space. In a very real sense, just as we might say that "matter" arises when the background "field strength" of the pure field becomes intense enough, so does the light of self-aware consciousness arise from dark background field consciousness when its density or strength is strong enough. The presence of dark pigmentation in neuromelanin appears to be crucial here, in that both the pre-synaptic and post-synaptic neural structures have "dense projections of gray," between which energy flows at the synaptic junctures. Whether this energy is in the form of electron tunneling or quantum "resonance" of some kind is an open question.

Our own intuition here supports a rhythmic and harmonic resonance process. Regardless of which model of "energy" and information transfer you perceive, the "quantum sea of light energy" that is both matter-energy and the very potential itself of matter-energy that pervades the universe, is actualized here from the boundless background reality that gives rise to space, time, matter, and all their interpenetrating, multidimensional dynamics. This is a vibratory universe," as above, so below," and so the warm dark matter of the brain inits affinity for shapes, forms, and processes must surely reflect in some way the patterns in the unseen cold dark matter of the universe. After all, the brain is a product and therefore in some ways a reflection of the structures of the universe, and so it is reasonable to locate some of these subtle cosmic structures and processes, with their innumerable convolutions and interconnections of space, time, and distances, in the structures and processes of the brain.

Melanin and neuromelanin have clearly established bioelectrical conductivity properties and may, under restricted conditions, manifest the phenomenon of biological superconductivity. In this context, research has established that the various mid-brain limbic system structures, e.g., amygdala, hippocampus, and inferior temporal lobe, are implicated in the stimulation/modulation of our primary emotions, ranging from rage to love to spiritual intoxication (Joseph, 2000). The neocortical surface of both the amygdala and the inferior temporal lobe have "dense neural fields" that recognize and respond to geometrical shapes and emotional imagery. These neurons are often called "feature detectors." This is both in response to external and internal, or physiognomic, perception. See Graph 1.

It is our suggestion here that neuromelanin provides these " pathways" of internal perception. There are also certain other quantum mechanical entrainment effects of cortical melanin with the wider solar, geogravitational, and electromagnetic environment that become focused under specific psychophysiological and cognitive conditions explored in diverse meditative disciplines alluded to earlier (Bynum, in progress). When these disciplines are consciously initiated in the attention association areas of the prefrontal cortex, they eventually affect the orientation association areas of the brain located in the posterior section of the left and right parietal lobes by what is termed "deafferentation," or the decrease of neural input (Newberg, D' Aquili, and Rause, 2002). This gray neuromelanin dense area is located in the top rear section of the brain in what is termed the posterior superior parietal lobe. The sense of space, time, self, and psychological boundary limitation is radically affected, and thereby the sensation of "travel," or shifting in space, time, and distance, is altered. This freeing of the self "reflection" from the conventional structures of space, time, and matter allows this self-sense to interact with other neural structures and processes to unfold an atypical but profoundly realized experience.

There is a neuromelanin nerve track strategically embedded in the brain-stem and "projected" upward through the mid-brain limbic system explored later in this text by Richard D. King. This Amenta nerve tract is concentrated most significantly in the substantia nigra at the beginning of a kind of loop in the track, and then it" rises up," first to the nucleus brachialis pigmentosus, then to the nucleus paranigralis, and finally to the locus coeruleus before descending in a long column that extends the length of the brainstem. See Chart III. "The column begins in the mid-brain, ventral to the somatic motor neurons of the third nerve nucleus, dorsomedial to the substantia nigra, and in contiguity to the nucleus paranigralis. [It ends] at the nucleus retroambigualis. In the remaining sections of the medulla, the column continues moving laterally until it ends in direct continuity with the intermediolateral gray of the cervical spinal cord"(Bazelon and Fenichel, 1967). When we observe this structure, we see that the "gray" of this brain constellation then moves down through the brainstem and projects directly down and into the spinal cord itself, forming a continuous gray melanin structure interconnected from beginning to end.

Chart III

Recent mapping of the human brainstem has located 12 areas with high concentrations of pigmented cells (Olszewski, 1964; Olszewski and Baxter, 1954). They are localized around the midline structures near the third and fourth ventricles, or cavities. They lie between the brain and its peripheral organs, and each communicates in complex ways with the cerebrospinal fluid.

1. Substantia nigra	7. Nucleus nervi Trigemini mesencephalic
2. Nucleus brachialis pigmentosus	8. Nucleus pontis centralis oralis
3. Nucleus paranigralis	9. Nucleus tegmental pedunculopontine
4. Locus coeruleus medialis	10. Nucleus parabrachialis
5. Nucleus intracapularis	11. Dorsomotor nucleus of thevagus
6. Nucleus subcoeruleus	12. Nucleus retroambigualis

Neuromelanin is exceedingly sensitive and receptive to light or luminosity in various modes. Melanin, in its affinity to certain laser characteristics, actually absorbs light, and these light absorbing melanin pigment vesicles possess both free radical-redox and ion control mechanisms along with the capacity for energy exchange or transformation from state to state by way of photon phonon transfer processes. It literally reflects an internally perceived bioluminosity. This is true for human beings in diverse cultures and meditative traditions across time.

In that context, we find here a parallel or complimentarity between an external light or image, reflecting some internal "physiognomic"perception of a "current" of inner light that is collectively projected out, and its expression mythologically in many ancient cultures and traditions, one example being the symbol of the "all seeing eye of Tibet." This projection outward of an internally perceived process is similar to what we have called "physiognomic perception" in developmental psychology (Werner, 1948; Werner and Kaplan, 1963).

This awareness seems to have reached a clinical level in the work of the Kemetic Egyptians who were astute medical observers of such phenomena and their relationship to illness (King, 1990). Their term was the "eye" of Horus. The Taoist medical practitioners developed procedures for the "circulation" of this blissful "eye," or light, in major and minor" orbits" through the body along the spinal line. Historically, both of these disciplines and traditions have coupled these energetic phenomena with other processes in a wider, more nonlocal solar ecology.

These disciplines have operational capacities in the modern world. The eminent anthropologists Marcel Griaule and Germaine Dieterlen (1986) report from the field that the Dogon of Mali, who trace their historical and genetic lineage to pre-dynastic Egypt some 5,000 years ago, by using a very similar methodology dating back over 600 years to at least the 13th century, were able to locate and map, without the aid of telescopes, the dwarf companion star to the star Sirius. The Dogon described its orbital "shape" and duration with percision. They described how it orbits on its own axis and also recognized its unique gravitational characteristics as a very small but dense star. This star, called Sirius B, or Digitaria, has a magnitude of 8 and is invisible to the naked eye. It makes a revolution around Sirius every 50 years, affecting the "shape" of its orbit and possessing elemental contents that mark it as a dense star. It was only first seen by telescope in the "modern scientific world" by two American astronomers, the Alwan Clarks, a father and son team, in 1862 and not actually photographed until

1970 by Irving Lindenblad of the U.S. Naval Observatory. The Dogon also accurately described another flashing object near Sirius B, only recently been seen by the NASA Einstein X-ray satellite, which turned out to be a dwarf nova.

The star Sirius is termed "sigi tolo" by the Dogon, and Digitaria is referred to as "po tolo." Digitaria, or po tolo, is actually a white dwarf or "embryological" star and is in the constellation of Orion. Sirius and its other satellites are called the "ku tolo," or" stars of the head," while the others are referred to as "gozu tolo," or" stars of the body." The Kemetic Egyptians referred to po tolo, or Digitaria, as the "sun behind the sun" and represented it as the hawk god Horus, enfolding the head of the pharaoh in their statues.

Interestingly enough from our perspective on earth, identification with the position of po tolo reveals that the world is turning as through on a *spiral*. There is a spiraling, elliptical shape in the orbits of Sirius and Digitaria, and there is a spiraling, serpentine shape in the modulating spatial structure between the first and last of the brain neuromelanin foci we are discussing. In other words, there appears to be a configurational similarity in the "orbits" of the outerworld and the inner world. "As above, so below. As within, so without."

Now, whether we accept these field reports by Griaule and Dieterlen or not, they express a controversial and contentious debate within science, with arguments claiming complete authenticity to others suggesting gross misinterpretation by these Western scientists. These Dogon claims might be dismissed if there were not such a long tradition of other such reports by the ancient scholars and astronomers from Proclus across a whole spectrum of eras and cultures, the earliest recorded being the Hermes Trismegistic literature, where it was referred to as the" Dark Mystery" or the" Black Rite" (Mead, 1964; Temple, 1976). This ancient part of the world records other accurate astronomical observations ranging from the pre-Christian megalithic site at Namoratunga (Lynch and

Robbins, 1978) in modern Kenya to the celestial ruins of Nabta Playa eons ago in the distant Nubian Sahara(Brophy, 2002).

Such claims often seem to rest on observations of the external world and correlations with the intimate functions of the human body. This suggests a deeper connection. It is quite possible that our current laws of physics, upon which we implicitly define our concepts and boundaries of mind, do not entirely confine the limits of consciousness. Future discoveries may greatly expand our understanding of this "above-below" connection. Our deepest intuition of consciousness will always be intimately associated with our most adventurous conception of light. These long-standing and consensually validated observations from the field by noted anthropologists, archeologists and astronomers, may be due to mere coincidence or chance. However, this now seems more and more unlikely. There is another possibility, perhaps another epistemology.

Does every tissue of space-time have enfolded within it inseed form the history and structure of the entire cosmos? As it unfolds, would each level or plane of explication in turn manifest a projection, amplification, and reverberation of these other progressively moreenfolded levels, surfaces, and orders? Within the body itself are there planes or surfaces, such as the palms of the hands, bottoms of the feet, the ear, tongue, or iris of the eyes, where the entire organization of the body can be found and "awakened"? After all, there is a bodily representation or sensory homunculus in the sensory cortex of the precentral gyms and a motor homunculus in the anterior central gyms on the brain's surface, as well as a form of motor homunculus configured lower in the mid-brain cerebellum region near the pons. In this interconnected and interpenetrating field of mind, body, and brain does each organ system manifest a signature vibratory mode amenable to conscious intervention under certain states of focused observation? And finally, within the intimate convolutions of the brain itself do these vortices and planes of manifestation, like the dynamics of gravity, provide latticework, pathways, and extended connectivity to the interwoven loom of space?

It is quite possible that the similarity of these inner and outer orbits in the brain to the physical universe, which we actually feel in states of unitative conscious experience, is due to a similarity awakened through a nonlocal connection, such as information exchange or travel through electron tunneling, or by way of a vibratory affinity, such as quantum resonance between these brain structures and the "representative space" these stars and other structures move through. In other words, there appears to be a hidden algorithmic contiguity between internal and external space created by an interaction between these neuromelanin foci, brain structures in vibration, and the topological curvatures of enfolded space.

In any event, these observations made by the Dogon and celebrated in their ceremonies have stood the test of time and are consensually "validated" by their religious disciplines and recently by modern astronomical science. Regardless of these ancient observations, this orbit or circulation or directly observable current of sensation, appears at the very least to reflect this nigrostriatial area of neuromelanin foci in the brain-stem that reverberates upward through the mid-brain limbic system and perhaps reflects the an lage pathway of the brainstem and below that emerged early in embryogenesis through the neural crest guided by the melanin and neuromelanin self-regulating process (Barr, 1983).

It is suggested that when the neuromelanin or nigrostriatial area is only mildly" activated," as in the procedure outlined in this chapter, the internal perception is of a subtle and relaxing luminous circuit or orbit along the spinal line that parallels the movement or neuromodulation along this column. When this nigrostriatial column however is fully "awakened" by diverse means and disciplines, there is a distinct and intense perception of bioluminosity emerging from this dark matter of the spine up into the braincore and beyond. It is subjectively experienced in the early stages along the spine as an undulating and living force. This may be due to the further resonant and rhythmic entrainment of neuromelanin foci, topologically situated in the third and fourth ventricles of the brain. The other organs in this dark inner chamber are termed the circumventricular organs. This level, or "plane," then, by a resonate affinity established by the properties of neuromelanin, would in turn further project, amplify, and reverberate up and into the region of the sensory motor cortex in the precentral gyrus, partly accounting for that peculiar circular sensory cortex "current" so often observed in classical meditative experiences (Bentov, 1977; Sannella, 1987).

It is significant to note for our purposes that these circumventricular organs are midline brain structures that border the third and fourth ventricles and are outside the blood-brain barrier. Because blood and the cerebrospinal fluid flow between structures and fluids more freely here than in other regions of the body and brain, there is a radical increase in communication between these structures, peripheral organs, and the blood-born products and information at these sites. These circumventricular organs include the pineal gland, median eminence, subfornical organ, area postrema, subcommissural organ, the organum vasculosum of the lamina terminal is, and also include the intermediate and neural lobes of the pituitary.

When these structures are set in motion by orchestrated breathing and concentration and/or ritual dance and coordinated bodily pulsation or other meditative disciplines, there is a synchronized vibration between the third and the fourth ventricles, which are connected by a tunnel of cerebrospinal fluid. When this vibratory spiral reaches the pineal gland, it stimulates it in an upward fashion. The pineal is attached to areas located in the fourth ventricle and floats in this vibrating pool of cerebrospinal fluid. Because neuromelanin has the capacity to transform energy from one state to another, i.e., the photon-phonon transfer process that occurs in apiezoelectric gel where mechanical vibrations are converted into electrical energy, it may supply the energy of this internally perceived "current" that appears to move along in the form of a staff or lower portion of a cross. The stimulated pineal would thereby become the newer and higher plane amplification and reverberation of this initial loop at the top of this perceived column or staff, and the diagonal or plane of this physiognomic perceived cross presumably would be that area where neuromelanin is most concentrated on the twelve-foci tract. This new loop of awakened energy would come to be symbolized as an "eye" or bird or other symbol of light, flight, insight, or illumination and freedom.

See Chart III and Graph 3.

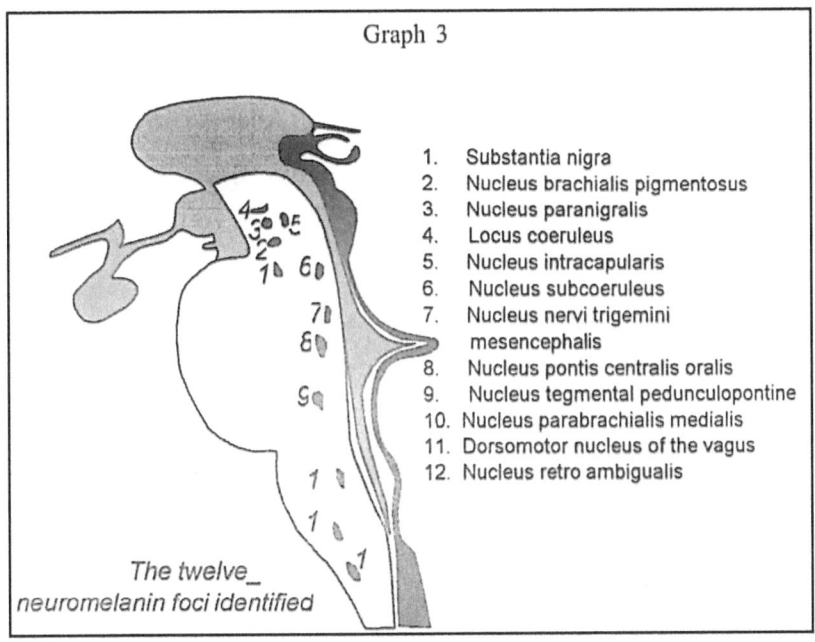

Graph 3

1. Substantia nigra
2. Nucleus brachialis pigmentosus
3. Nucleus paranigralis
4. Locus coeruleus
5. Nucleus intracapularis
6. Nucleus subcoeruleus
7. Nucleus nervi trigemini mesencephalis
8. Nucleus pontis centralis oralis
9. Nucleus tegmental pedunculopontine
10. Nucleus parabrachialis medialis
11. Dorsomotor nucleus of the vagus
12. Nucleus retro ambigualis

The twelve neuromelanin foci identified

In the later stages, when the meditative process has been experienced and fully realized by enough practitioners, the dense "feature detecting" neurons of the limbic system, especially the neocortical surface of the inferior temporal lobe and the amygdala (Joseph, 2000), would project outward this luminous closed loop or elliptical cross or ankh that many others would come to recognize as a primordial human process (Bynum, 1993). This light-interacting neuromelanin web and its projections, beginning early in embryo genesis and continuing on through the subsequent stages of development, provide for the latticework of other more subtle bioelectrical pathways through the higher, enfolded cerebral structures of the brain, known in the disciplines of meditative reflection but as yet not within the province of biomolecular science (Bynum, in progress).

Again, piezoelectric and other inherent phenomena of biological organisms, such as electro-streaming properties, provide for the transduction of energy and information from movement, mechanical stresses, vibration, and sound into electromagnetic oscillations. Magnetic fields in particular penetrate through the body and skull with relatively little loss of amplitude and provide for dynamic electrical changes in the process.

For human beings, the pineal gland in particular functions not only as a site of photosensitive processes but its organ-design function and strategic location in a vibrating fluid sea allow it to be an EMF (electromagnetic field) sensor for these magnetic fields and to detect ELF (extremely low frequency) oscillations. These are directly experienced in meditative states and other conditions of focused subjective attention.

The highly ordered water and cellular state of living matter, a nearly crystalline latticework or structure, versus the significantly less well-ordered state of nonliving systems, makes biological superconductivity (type II) a possibility at room temperature as opposed to the kind of superconductivity (type I) that is observed at extremely low temperatures in nonliving systems. Electron pairs flow more freely across small junctures at these quantum mechanical levels of enfolded space so that the usual "space distance" dimension is less crucial for the communication of infomlation than the "form-resonance" or quantum mechanical tunneling dynamism.

A crucial aspect of this type II, or biological, superconductivity is that it transmits the magnetic field in multiples or rhythms of unit quantum magnetic flux, thereby making the organism potentially sensitive, under certain restricted and disciplined conditions, to extremely subtle magnetic and gravitational fields (Dubrov, 1978). Remember that of the four fundamental forces currently identified in physical nature, it is only electromagnetism and gravity that are actually experienced in human consciousness. The so-called "strong force" and "weak force" are confined to the internal structure of the atom.

It is known that vast geothermal and magnetic forces within the earth's rotating core dynamics torque upward and generate an electromagnetic field in discrete lines of force over the surface of the earth. Einstein showed that gravity itself can be understood as an expression of the geometry of space and is woven into the loom of space-time. Also, we remember that electromagnetism and gravity are both experienced through the nexus of human consciousness. It is therefore not unreasonable to include the subtle effect of both internally generated terrestrial magnetism, gravity and electromagnetism, and external astronomical gravitational fields within this wider topologically complex ecology.

A quieted and sensitive nervous system can be made aware of this matrix of subtle but dynamic relationships, both as they emerge from within the earth and as they descend into the earth through subtle lines of force, along which they are also conducted. In states of deep meditation, healing, or other contemplative disciplines, when there is a resonance between the base heart aorta rhythm (roughly 7 hz), the earth-ionosphere electromagnetic cavity radiance at multiples of 7.8 hz, termed the Schumann resonance, and wider solar entrainment effects, we enter into unusual states of expanded awareness (Bynum, in progress). Under such restricted conditions, psychologically speaking, the loom of space-time itself appears as mutable as a conscious dream.

The bioelectrical properties of neuromelanin and melanin in the brain and along the brainstem have recently begun to receive a great deal of attention in the clinical literature (McGinness, Corry, and Proctor, 1974; McGinness and Proctor, 1973; Barr, 1983). Melanin and neuromelanin, already clearly established as light sensitive semiconductors, are excellent candidates for the role of neuromodulator of the central nervous system (Longue-Higgins, 1960; Filatous, McGinness, and Corry, 1976; Filatous, McGinness, and Williams, 1980). Both subtle magnetic fields and their associated bioelectrical currents can be stimulated in the brain, which then profoundly affects human consciousness. As this is more and more born out in clinical practice, it will revolutionize the life sciences. It brings into possibility the effect of biological superconductivity, a relatively new hypothesis in clinical science (Little, 1965; Cope, 1979 and 1981). One immediately thinks not only of the psychophysiological symptoms already mentioned but also of symptoms such as seasonal affective disorder (Terman, Terman, Schlager, et al., 1990; Light for Better Living) and other depressive symptoms in addition to symptoms involving decreased subjective energy or underactivation of the autonomic nervous system. In terms of understanding the diverse meditative disciplines of the earth's peoples and their connectivity to the wider terrestrialand cosmic ambience, its capacity for unification and a collective psychospiritual trajectory is beyond our present comprehension.

CONCLUSION

We have presented a standardized technique for activation of the autonomic nervous system, which appears to be intimately associated with the psychophysical and neurodynamics of neuromelanin. This technique can be taught in a standardized way to patients responding to or suffering from acute or prolonged psychophysiological symptomatology. It involves definite and well known psychophysical and subjective reactions. However, beyond these primarily therapeutic functions, it may also be indicated inclinical and subtle psychophysiological and bioelectrical or superconductive processes, which we are only beginning to understand.

There even appear to be nonlocal relationships and perhaps topological contiguities between microneural processes and the macroprocesses of the cosmos, specifically a configurational similarity between certain brain neuromelanin foci and spiraling constellational structures. We suspect that, under restricted conditions, consciousness may share with quantum mechanics certain general features, such as the capacity for trans-temporal and trans-spatial information exchange, or what is termed nonlocality. These are events that occur within the universe of the body as well as the wider physical universe with which the body interfaces, both being interpenetrated by the same principles thatintimately connect us all in a more inclusive ecology.

What applies to one above or "within" applies to the other below or "without." We look farther down the road toward this clinical procedure that allows us to communicate more directly and clearly with the body. It offers a great deal of ope to the patient suffering from seemingly intractable pain and illnesses.

REFERENCES

Achterberg, J., and Lawlis, G. F., 1984. *Imagery and Disease.* Champaign, IL: LP.A.T.

"Alternating Cerebral Hemispheric Activity and Lateralization of Autonomic Nervous Function," *Human Neurobiology,* 39-43.

Amaral, D. G., and Sinnamon, H. M., 1977. "Locus Coeruleus. Neurobiology of a Central Noradrenergic Nucleus. Progress in Neurobiology,"vol. 9, 147-196.

Bandler, R., and Grinder, J., 1982. *Reframing: Neuro-Linguistic Programming and the Transformation of Meaning.* Moab, UT: Real People Press.

Barr, E. F., 1983. "Melanin: The Organizing Molecule," *Medical Hypotheses,* 11, 1, 1-139.

Bazelon, M., and Fenichel, G. M., 1967. "Studies on Neuromelanin.
A Melanin System in the Human Adult Brainstem," *Neurology,* 17, 512-519.

Bentov, I., 1977. *Stalking the Wild Pendulum: On the Mechanics of Consciousness.* New York: Dutton.

Bogerts, B., 1981. "A Brainstem Atlas of Catecholaminergic Neurons *in Man,Using Melanin as a Natural Marker,"Journal of Comparative Neurology,* 197, 63-80.

Brophy, T.G., 2002. *The Origin Map: Discovery of a Prehistoric, Megalithic, Astrophysical Map and Sculpture of the Universe.* New York, Lincoln, Shanghai: Writers Club Press.

Bynum, E. B., (2012). Dark Light Consciousness. Inner Traditions and Bear Company, Rochester, VT.

-1993.*Transcending Psychoneurotic Disturbances.* Binghamton, NY: Haworth Press.

-in progress. *Kundalini and the Psychobiology of Transcendence.*

Charney, D.S., and Heninger, G. R., 1986. "Abnormal Regulation of Noradrenergic Function in Panic Disorders," *Archives of General Psychiatry,* 43, 1042-1054.

Charney, D.S., Heninger, G. R., andBreier,A., 1984. "Noradrenergic Function in Panic Anxiety," *Archives of General Psychiatry,* 41, 751-763.

Cope, F. W., 1979. "Remnant Magnetization in Biological Materials and Systems as Evidence for Possible Superconductivity at Room Temperature: A Preliminary Survey," *Physiological Chemistry and Physics,* 11, 65-69.

-1981. "Organic Superconductive Phenomena...," *Physiological Chemistry and Physics,* 13, 99-110.

Cotzias, G. C., 1974. "Melanogenesis and Extra Pyramidal Diseases, *Fed. Proc.,* 23, 713.

Diop,A. C., 1974. *The African Origin of Civilization,* M. Cook, trans. Westport, CT: Lawrence Hill and Co.

-1991. *Civilization or Barbarism: An Authentic Anthropology,* Yaa-Lengi Meema Ngemi, trans. Brooklyn, NY: Lawrence Hill Books.

Dubrov, A. P., 1978. *The Geomagnetic Field and Life: Geomagneto biology,* F. L. Sinclair, trans. New York: Plenum.

Ebbell, B., 1937. *The Papyrus Ebers: The Greatest Egyptian Medical Document,* B. Ebbell, trans. Copenhagen, Levin and Munksgaard.

Edmonston, W. E., 1986. *The Induction of Hypnosis.* New York: J. Wiley and Sons.

Elama, M., Svensson, T. H. E., and Thoren, P., 1986. "Locus Coeruleus Neurons and Sympathetic Nerves: Activation by

Visceral Afferents," *Brain Research,* 375, 117-125.

Erickson, M. H., and Rossi, E. L., 1979. *Hypnotherapy: An Exploratory Casebook.* New York: Irvington Publishers, 94-142.

Fenichel, G. M., and Bazelon, M., 1968. "Studies on Neuromelanin. Melanin in the Brainstem of Infants and Children," *Neurology,* 18, 817-820.

Filatous, G. J., McGuinness, J.E., and Williams, L., 1980. "Statistical Analysis of Switching in Melanin," *Physiological Chemistry and Physics,* 12, 534-538.

Filatous, J., McGinness, J.. and Corry, P., 1976. "Thermal and Electronic Contiibutions to Switching in Melanins," *Biopolymers,* 15, 2309-2319.

Finch, C. S., 1990. *African Background to Medical Science.* London: Karnak House.

Foote, S. L., Bloom, F. E., and Aston-Jones, C., 1983. "Nucleus Locus Coeruleus: New Evidence of Anatomical and Physiological Specificity," *Physiological Review,* 63, 844-914.

Fried, R., 1987. *The Hyperventilation Syndrome: Research and Clinical Treatment.* Baltimore and London: Johns Hopkins University Press.

Funderbunk, J., 1977. *Science Studies Yoga: A Review of Physiological Data.* Honesdale, PA: Himalayan International Institute of Yoga Science and Philosophy.

Gillis, R.A., Quest, J.A., Pagani, F. D., etal., 1991. "Control Centers in the Central Nervous System for Regulating Gastrointestinal Motility, in R. A. Gillis, J. A. Quest, and F. D. Pagani, eds., *Handbook of Physiology.* New York: Oxford University Press.

Gorman, J.M., Liebowitz, M. R., Fyer,A. J., etal., 1989. "A Neuro anatomical Hypothesis for Panic Disorder," *American*

Journal of Psychiatry, 146, 148-161.
Griaule, M., andDieterlen, G., 1986. *The Pale Fox.* Az. Continuum Foundation.

Grinder, L., and Bandier, R., 1981. *Trance-formation: Neuro Linguistic Programming and the Structure of Hypnosis.* Moab, UT: Real People Press.

Hassert, D. L., Miyashita, T., and Williams, C. L., 2004. "The Effects of Peripheral Vagal Nerve Stimulation ...," *Behavioral Neuroscience,* vol. 1,118.

Hourning, E., 1986. "The Discovery of the Unconscious in Ancient Egypt," *An Annual of Archetypal Psychology and Jungian Thought,* 16, 28.

Huang Ti, 1966. *The Yellow Emperors Classic of Internal Medicine,* I. Veith, trans. Berkeley, CA: University of California Press.

Jackson, J. G., 1970. *Introduction to African Civilization.* Secaucus, NJ: Citadel Press.

Joseph, R., 2000. *The Transmitter to God: The Limbic System, the Soul and Spirituality.* San Jose, CA: University of California Press.

Kagan, J., and Rosenberg, A., 1987. "Iris Pigmentation and Behavioral Inhibition," *Developmental Psychobiology,* 20, 377-392.

Kiecolt-Glaser, J. D., Stephens, R. E., et al., 1985. "Distress and DNA Repair in Human Lymphocytes," *Journal of Behavioral Medicine,* 8, 311-320.

King, R. D., 1990. *The African Origin of Biological Psychiatry.* Germantown, TN: Seymore Smith, Inc.

-1990. "The Question of Melanin-The Study of Blackness" (video). Washington, DC: National Association of Black Psychologists.

Klein, R., and Armitage, R., 1979. "Rhythms in Human Performance: 1 1/2 Hour Oscillations in Cognitive Style," *Science,* 204, 1326-1328.

Kuvalayananda,S., 1978. *Pranayama.* Philadelphia: Sky Foundation.

Light for Better Living: The Ultra-Bright Light Systems. Medic Light, Inc., Yacht Club Drive, Lake Hopatcong, NJ 07849 Little, W.A., 1965. "Superconductivity At Room Temperature," *Scientific American,* 212, 21-27.

Locke, S., and Colligan, D., 1986. *The Healer Within: The New Medicine of Mind and Body.* Bergenfield, NJ: New American Library.

Longue-Higgins, H. C., 1960. "On the Origin of the Free Radical Property of Melanins," *Archives of Biochemistry and Biophysics,* 86, 231-232.

Lynch, B. M. and Robbins, L. H., 1978. "Namoratunga: The First Archaeoastronomical evidence in sub-saharanA:frica." *Science,* 200, 766-768.

Lysebeth,A. V, 1983. *Pranayama: The YogaofBreathing.* London, Boson, and Sydney: Unwin Paperbacks.

Marsden, C. D., 1961. "Pigmentation in the Nucleum Substantiae Nigrae of Mammals," *Journal of Anatomy,* 95, 162-256.

McGinness, J., and Proctor, P., 1973. "The Importance of the Fact That Melanin Is Black," *Journal of Theoretical Biology,* 39, 677-678.

McGinness, J., Corry, P., and Proctor, P., 1974. "Amorphous Semiconductor Switching in Melanins." *Science,* 183, 853-855.

Mead, G. R. S, 1964. *Thrice Greatest Hermes.* London:

John Watkins Publishers

Meyer, J. S., 1985. "Biochemical Effects of Corticosteroids on Neural Tissue," *Physiological Review,* 65, 946-1020.

Moore, T. O., 1995. *The Science of Melanin: Dispelling the Myths.* Silver Spring, M.D.: Beckham House.

-2002. *Dark Matters, Dark Secrets.* Redan GA: Zamani Press.

Motoyama, H., 1981. *Theories of the Chakras: Bridge to Higher Consciousness,* 142-154. Wheaton, IL: Theosophical Publishing House.

Muses, C., 1972. "Trance-Induction Techniques in Ancient Egypt," in C. Muses and A. M. Young, eds., *Consciousness and Reality: The Human Pivot Point,* 9-17. New York: Avon Books.

Netter, F. H., 1972. *The CIBA Collection of Medical Illustrations,* vol. 1, "The Nervous System." Summit, NJ: CIBA Press.

Newberg, A., D' Aquili, E., and Rause, V., 2002. *Why God Won't Go Away.* New York: Ballantine Books.

Olszewski, J., 1964. *Cytoarchitecture of the Human Brain.* New York: Stemand Birjelow.

Olszewski, J., and Baxter, D., 1954. *Cytoarchitecture of the Human Brain Stem.* Basel and New York: S. Karger.

Prakashan, P., 1980. *Shiva Svarodaya.* Varanasi, India: Chowkhamba Sanskrit Series.

Rama, S., 1981. "Application of Sushumna" (Cassette Recording No. 0308). Honesdale, PA: Himalayan International Institute of Yoga Science and Philosophy.

Rama, S., Ballentine, R., and Hymes,A., 1979. *Science of Breath.* Honesdale, PA: Himalayan International Institute of

Yoga Science and Philosophy.
Rider, M. S., Achterberg, J., Lawlis, G. F., et al., 1990. "Effect of Immune System Imagery on Secretory IgA," *Bio feedback and Self-Regulation,* 15, 317-333.

Sannella, L., 1987. *The Kundalini Experience: Psychosis of Transcendence.* Lower Lake, CA: Integral Publishing.

Scherer, H.J., 1939. "Melanin Pigmentation of the Substantia Nigra in Primates," *Journal of Comparative Anatomy, 71,* 91-95.

Siever, I. J., Uhde, T. W., Jimerson, D. C., et al., 1984. "Differential Inhibitory Noradrenergic Responses...," *American Journal of Psychiatry,* 141, 733-741.

Sperry, R. W., 1988. "Psychology's Mentalist Paradigm and the Religion/Science Tension," *American Psychologist,* August, 607-612.

Svatmarama, S., 1971. "The Shambhavi Mudra and the Inner Light," in *The Yoga of Light: Hatha Yoga Pradipika,* 164-175, H. Rieker and E. Becherer, trans. Middletown, CA: Dawn House Press.

Temple, R. K. G., 1976. *The Sirius Mystery.* London: Sidgwick & Jackson.

Terman, J. S., Terman, M., Schlager, D., et al., 1990. "Efficacy of Brief, Intense Light Exposure for Treatment of Winter Depression," *Psychopharmacology Bulletin,* 26, 3-10.

Thakkur, C. G., 1977. *Yoga: Harmony of Body, Mind and Soul.*
Bombay, India: Yoga Research Center/AncientWisdom Publications.

Tokay, E., 1972. *Fundamentals of Physiology.* New York: Barnes and Noble.

Warner, H., and Kaplan, B., 1963. *Symbol Formation.* New York:

J. Wiley and Sons.

Werner, H. 1948. *Comparative Psychology of Mental Development.* New York: International Universities Press.

Werntz, D., 1981. *Cerebral Hemisphere Activity and Autonomic Nervous Function.* Unpublished doctoral dissertation, University of California, San Diego.

Werntz, D., Bickford, R., Bloom, F., and Shannahoff-Khalsa, D., 1981. "Selective Cortical Activation by Alternating Autonomic Function." Paper presented at the Western EEG Society Meeting, Reno, Nevada.

The Clinical Use of Bliss:

NOTES

The Clinical Use of Bliss:

NOTES

CHAPTER 4

Neuromelanin: A Black Gate Threshold; The I-33 Tissue Of Heru, Historical, Neurophysiological, And Clinical Psychological Issues
Richard D. King, M.D.

INTRODUCTION

"O Wosir, the King, take the Eye of the living Heru that you may see with it. O Wosir, the King, may your vision be cleared by means of the light. O Wosir, the King, may your vision be brightened by the dawn. O Wosir, the King, I give to you the Eye of Heru when Ra gives it. O Wosir, the King, I put forth to you the Eye of Heru on that you may see with it."

-Utterance 639 (Faulkner, 1969), Pyramid Texts, 3200-2100 B.C.E., Kemetic Old Kingdom

SCOPE, METHODS, AND ISSUES

This chapter will review the subject of neuromelanin for several issues that are related to (1) the history of the ancient African study of Black symbolism/melanin/neuromelanin; (2) the current era's scientific study of concepts related to neuromelanin; (3) the role of neurogenesis or new nerves as they are added to the cortex and affected by psychosocial stressors; and (4) current era concepts of the clinical practice of psychology/psychiatry related to neuromelanin anatomy and physiology.

The methodology is one of clinical psychiatry, which focuses on the use of dynamic symbolism, ontological meaning, and emotional-historical context in the analysis and understanding of human motivation, behavior, and consciousness. Dynamic symbolism is a reflection of internal processes and has the capacity to transform and heal the body and mind when intelligently focused and understood. It is therefore critical in this light that a study of neuromelanin be seen as a study that is an integral part of a much larger study of melanin, not only skin melanin but brain melanin in the classical and historical tradition of melanin as the Kemetic flesh of Ra (Piankfoff, 1954).

The neurocosmological implications of the Black carbon atom and especially Black cosmic melanin/nanodiamonds of carbon-rich proto-planetary nebulae interstellar Black matter is also implicated in this wider view that embraces both the micro worlds within us and the cosmic ambience that dwells about us. For as surely as this is one universe, the inner and outer aspects of it must interpenetrate and meet at some level of manifestation. This is indeed the whole thrust of the ancient vision on the banks of the Nile in Kemet and the root of our modem concept of it as neurocosmology (Van Kerekhoven, 2002; Hill, 1998; Luu, 1993; Bradley, 1966; Liou 1996; Nicolaus, 1998; B. Nicolaus, 1997; R. Nicolaus, 1997).

MELANIN, NANODIAMONDS, AND THE LINEAGE OF PANSPERMIA

The cosmic carbon atom interstellar gas clouds of nanodiamonds/cosmic melanin and their relationships to biological melanin and neuromelanin is a fundamental category of knowledge that yields a critical meaning to a study of melanin which dismisses the two tendencies in modem science that seek for ideological reasons to lessen the impact of melanin studies. One tendency is to dismiss melanin studies simply as pseudoscience. This is to equate and associate melanin with the mere waste products of cellular metabolism. Is melanin a biological waste product or a luminous jewel waiting to be rediscovered and understood by us in modem times (B. Nicolaus, 1998)? Clearly, the absurdity of such a devaluation of melanin studies is made readily apparent by the intuitive knowledge that interstellar Black matter is in the form of interstellar gas clouds laden with these luminous jewels of Black nanodiamonds, which are the literal genetic seeds from which stars are born. They arise from the same high-energy primordial shock and flux that populated all of creation.

In this chapter we are openly suggesting that there truly is a crucial connection between the Black, or dark, matter that structures much of the unseen universe and the subtle living dark matter that structures our very bodies, brains, and nervous systems. The study of higher mathematics, physics, and chemistry, which observes such cosmic melanin macrocosmic phenomena, should be a wake-up call that a similar higher science is required to study carbon/melanin related to microcosmic phenomena in living biosystems on planet Earth, i.e., humanity. As Professor Brown pointed out in Chapter 2 of this book, the cosmos itself is suffused with different forms of melanated matter, from the stellar to the planetary. Is inner melanin in some as yet unknown way a gateway to the outer world of dark matter, as suggested in the last chapter by Bynum? Where does this black material so intimately associated with life itself come from in the first place?

The interstellar material of our Milky Way galaxy contains in a huge state vast interstellar gas clouds composed of hydrogen (70%) and helium (28%), with a small component of solid particles and interstellar, or cosmic, dust. In addition to the element carbon, there is found in interstellar expanses the elements oxygen, nitrogen, nickel, sulfur, aluminum, iron, and others, along with many different organic and inorganic molecules. It is believed that following the initial so-called Big Bang of creation, the universe expanded in a hot burst of pure energy then later cooled down into a condensation of early subatomic quark soup, leading to the first element, hydrogen. Stars were formed by this process, and they were initially mostly hydrogen.

The process of gravitational collapse and other cosmic forces eventually created a nuclear fusion state that fused the nuclei of hydrogen atoms together to produce helium atoms. The helium then continued to fuse, producing atoms of progressively higher atomic number, nuclei with more protons and neutrons, and fusion, where internal heat gradually increased and transformed the structure of gas particles. It is out of this flux of "mother" gas clouds that the carbon atom arose. This is a vast simplification of a complex process, but the point is that the critical end product of black nanodiamonds is carbon. Van Kerekhoven (2002) reported that the temperature, pressure, and composition/molecular precursors in the solar nebula would favor the condensation of carbonaceous compounds, which are called nanodiamonds. Diamond formation is favored by an abundance of atomic hydrogen and low carbon ratios. Nanodiamonds are a common by-product of star formation and are formed in stellar systems and ejecta from a supernova carbon star explosion.

Meteorites containing interstellar diamonds were first reported in 1989 (Lewis, 1989). The extra-solar origin of the nanodiamonds was indicated by isotopic anomalies of 15n depletions, P and R Xe compositions, and enhancements of D/H ratios (Anders, 1993). Different models of the origin of meteoritic nanodiamonds have been reviewed (Van K, 2002; Anders, 1993), ranging from interstellar shocks (Tielens, 1987) to formations in exotic stellar locations (Jorgenson, 1988) to the quantum heating of carbonaceous grains (Nuth, 1992).

Nanodiamonds are a solid crystalline form of mostly carbon atoms that are extremely small in size, with median size of ~ 3nm (Lewis, 1989; Daulton, 1996). This is a size that is ten to one thousand times smaller than interstellar grains (Van Kerckhoven, 2002). Nanodiamonds have a large surface-area-to-volume ratio, with an active surface chemistry (Hill, 1998). It is the nanodiamond's active surface chemistry that results in the formation of active species such CH, CH2, CO, and NH (Hill, 1998). A critical role is served by nanodiamonds and interstellar gas clouds in the formation of stars and also in the creation of biogenetic molecules of melanin in the interstellar clouds of many galaxies and continues in the same way in our solar system, perhaps since the solar system was created. It is through the interstellar gas clouds that these black biogenetic surfaces, sometimes transported on the larger surfaces of traveling comets along with amino acids, moved through the stellar abyss and throughout the innumerable solar systems, seeding the surface of planets like our Earth. This is the ancient "panspermia" hypothesis and vision reinterpreted by Nobel laureates Francis Crick and Leslie Orgel (Jantsch, 1980). It is these Black matter melanin seeds from the stellar expanses that provide the biogenetic spark of life. From this perspective, it is difficult to see how this material could ever be conceived as mere "waste" matter.

SCIENCE, MEMORY, AND THE AMENTA

There is a second tendency that continues to lessen the scientific impact of melanin studies and that one equates high levels of human melanin skin content with racial inferiority (Jablonski, 2000; Kershaw, 1998; Kershaw, 1999; Guderian, 1952). Yet every clinician and social scientist of this century knows that over the last 500 years there has been a painful relationship between the devaluation of people with high levels of melanin in their surface skin pigmentation and the psychology of White supremacy/racism (Welsing, 1970; Welsing, 1990; Ani, 1994). Again, a firm grasp of advanced knowledge in the fields of psychology, neuroanatomy, neuroanatomical embryology, neuropharmacology, and historical epigenetic unfoldment are absolutely required to appreciate the deeper dynamic role of melanin in human experience.

Also to be considered are the PTSD (post-traumatic stress disorder) psychological emotional trigger effects of visual images of Black symbolism that evoke the outpouring of emotion-laden memory projections from within the human mind. These are the levels, or multiple gates, of the Kemetic Amenta, the superconscious, personal unconscious, lower unconscious, and collective unconscious, all of which are manifestations and projections of the ancient Kemetic understanding of the Primeval Waters of Nun (Assagioli, 1965; Bynum, 1984; Bynum, 1999; King, 1994).

Today's human mind is pregnant with millions, if not billions, of years of emotional, intense active memories of the whole spectrum of one's blood line genetic ancestors. These experiences form a literal ocean of incredibly beautiful human experiences that range from heaven to hell, from the heights of love, romance, and creative genius through the lows of fear, jealousy, and post-traumatic stress disorder (PTSD). This great phylogenetic storehouse of collective memory even includes our fragmented memories of exposure to geological catastrophes. This is an ocean of living memories of the tragic conversion of the lush water-filled grasslands of North Africa into today's largely barren Sahara Desert, Eurasian glacial ice ages, floods, meteor strikes, earthquakes, and volcanic eruptions, not to mention the countless wars of ethnic cleansing.

Amid these, however, are also those quiet islands of the triumphs of human cooperation, communication, and compassion in the interests of mutual survival and evolution toward some higher purpose and union. This has been an epic saga and struggle of hominid evolutionary consciousness from the early australopithecines of pre antiquity up to the Homo sapiens sapiens of today (Bynum, 1999).

Deeper yet, there is in this continuum of living history the current era's oppressive paradigm of White supremacy/racism that is born from a very great fear of melanin Blackness and of a melanin death/ sleep that ironically arises from the comical reality of a missed opportunity, a closed door, a tragic, aborted unfolding or non-opening of the psychospiritual door to the higher-order melanin-structured biophysical systems of luminous blackness.

The psychotic yet comical reality of White supremacy witnesses a failure of "growing up," a failure to evolve from a neophytic, strictly logical left-brain stage of consciousness to the second stage, a stage of intelligence with inner vision and dual cortical hemispheric integration, then further on to the third stage of sons/daughters of light, unity of light itself with a Christ-consciousness-level of inner vision, the perfected human incarnation of biophysical evolution. Thus, White supremacists and the victims of White supremacy experience a horrific tragedy of lowered standards for human potential, a failed adult human transformation in the alchemical bath of life in which the head of the Ethiopian is the Black philosopher's stone, the Anu Benben stone, melanin/neuromelanin I-33 brainstem spinal column, which is the threshold door through which such transformations are born.

There are melanin systems that transform the seemingly inert Black earth into the Black diamond of inner vision (King, 1993; Budge, 1967; Faulkner, 1967; Faulkner, 1978; Piankoff, 1977). Inner vision is an advanced science, a disciplined form of introspection, not a form of regression to pre-scientific thinking or pseudoscience. Humanity as yet is only in the anteroom of the next stage of our evolutionary drama. The obscuring Blackness holds the key to the hidden light of inner vision.

HISTORICAL ISSUES IN THE ANCIENT AFRICAN STUDY OF BLACK SYMBOLISM/MELANIN/ NEUROMELANIN

African scientists seeking a Black consciousness throughout the past several million years, if not countless ages, have deeply studied nature's script of Black symbolism (James, 1954; Diop, 1991; Clarke, 1999; Allen, 1974; Jackson, 1970; King, 1994; Faulkner, 1969; Faulkner, 1978; Churchward, 1913; Churchward, 1921). Skin melanin, which ancient Egyptians referred to as the "Flesh of Ra" (Piankoff, 1954) and "I-33 Tissue of Horus" (Piankoff, 1977), and neuromelanin have been a part of such studies. Neuromelanin is a Black biopigment that is present in neuro, or brain, tissue including the outer coverings of the brain, the dura mater and pia mater, and the substance of the brain neurons and glial cells.

This brief history of ancient African study of melanin/ neuromelanin will concern 5 issues: (1) the Kemetic root name, M3NW, for melanin; (2) the pre-dynastic Kemetic Memphite cosmology concept of the Hill (Ptah/mind); (3) Kemetic architectural symbolism of the Black Rose Granite Threshold stone; (4) Kemetic architectural symbolism of the Black Rose Granite all Black Kings chamber in the Great Pyramid at Ghiza; and (5) Kemetic statements of neuroanatomy from 1349 B.C. on the right panel second shrine in the Tomb of Tutankamen.

The history of the ancient African study of melanin begins with the name itself, melanin. The history of the name melanin is that it is an English word derived from the Greek word melan, which means black (Bernal, 1991). Bernal reported, "There is no common Indo-European root for the color black. .. however, it would seem more plausible to derive this from the Egyptian (Kemetic) name. M3NW, the mountain in the West, where the sun goes down in the evening, and enters the underworld."

The Egyptians, hereafter referred to as Kamites, inherited from Khuiland and later Ethiopian ancestors (King, 1990; 1994) the M3NW concept. The M3NW concept was part of a philosophical system that witnessed a daily, 24-hour cycle of daylight and darkness, in which the sun was seen at sunset to enter the earth at M3NW (melanin), the Western Mountain of the Moon, and undergo a 12-hour nighttime passage through the all-Black domain of Amenta, passing through 12 gates and then emerging at dawn through an eastern M3NW (melanin) gate, at the Eastern Mountain of the Moon for the 12-hour passage of the sun during daylight.

The extreme antiquity of millions of years of the M3NW root name for melanin is readily apparent. Consider the fact that the geographical place in Africa where the 24-hourday is equally divided into 12 hours of day and 12 hours of night is at the equator of the planet. Moreover, it is in the equatorial region that there are not only ancient African records of an Eastern Mountain of the Moon, BAKHU, Kenya's Mount Kilimanjaro (Ben Jochannan, 1972) and a Western Mountain on the Moon, MANA, the Renzori Mountain Range. In fact, Ben Jochannan (1972) cited the records of the papyrus of Hunefer that stated that "we Kamites came from the Mountain of the Moon (Kilimanjaro Ethiopians)." Moreover, there are abundant records of Egyptian references to the Great Lake region between the Eastern Mountain of the Moon and Western Mountain as the site of origin of River Hapi. This source of the Nile is the abode of the oldest Egyptian God, Bes, an Anu-Twa person from the Great Lakes region of Khuiland, with one group of Anu gathered around the Bakhu eastern mountains of the Moon and another group of Anu ancestors gathered around the Manu western mountain of the moon.

Certainly, it will require advanced studies in multiple fields, such as science, physics, electromagnetism, neurochemistry, and music, to answer the great scientific questions that our misperception and ignorance now mislabel as pseudoscience. We must explore the question of what special and perhaps unknown energetic-field relationships exist at the equator of this planet that promote epigenetic unfoldment of upright, walking biological systems called consciously evolving humans.

Budge has written (Budge, 1969) "Bes (the god Bes) has the same type of face as the Pygmy... the earliest mythology of old Egypt, and no doubt Bes was at a later date made to represent a type of Horus I (Jesus), who was at first their 'Chief of the Nome.' It was from these Anu (Pygmies) that the first mythology of Egypt sprung." The mythology was humanity's first use of the spoken word, symbolic body postures (signs, the seen word), rhythmical song/dance (the emotionally/harmonically expressed word), and later written records of deep, emotionally charged memories of the past experiences of bloodline genetic ancestors. Thus, current records have confirmed the common origin of all humanity from the same Khuiland Great Lakes region in northeast Africa before 7 million years ago and subsequent later migrations east into Ethiopia, north to Egypt and Eurasia, south to South Africa, and west to Chad, West Africa, and North Africa.

Current anthropological records confirm that the first 5 or more of the past 7 million years were spent in Africa, with only the last 2 million years seeing hominid ancestor migrations out of Africa, beginning with Homo erectus. The current branching of the hominid line shows that our own Homo sapiens sapiens species is only roughly 200,000 years old. This stock again originated in Africa and then migrated out of Africa onto every continent of the planet.

The Amenta that is remembered here is composed of that ocean of common multimillion-year experiences of our ancient common bloodline ancestors, the root bloodline of all humans, regardless of surface skin pigmentation. We are all Africans in our bones and genes (Bynum, 1999; Bruwet, 2002; Wood, 2000; Leakey, 1995; Kimbel, 1994; Kappelman, 1996; Asfaw, 1999; Tobias, 1987; Leigh, 1992; Falk, 2000; McHenry, 1998; Reed, 1997; Skelton, 1992). In fact, the Anu were the first humans to travel "into" the earth, the first humans to become conscious of the domains and doors within their own unconscious, and the first to map Amenta. In their mythology of the Travel of the Great Hero,

Ife, they developed a stellar mythos dating to 2,000,000 B.C.E. (Halet; King, 1992; Chruchward, 1913; Chruchward, 1924).

INVOLUTION AND EVOLUTION: THE DARK MATTER CYCLES OF SPIRIT AND COSMOGENESIS

According to James (James, 1954), there exist predynastic records from before the times of Dynastic Kemet (4000 B.C.). This was the Memphite Theology of the Primate of the Gods. Ptah, who first arose as a prominent hill from the water of Nun, is an island in the primeval lake in Khuiland (Chruchward, 1913), or multidimensional vibrational space. The primeval hill arising from the waters of Nun is a symbolic metaphor for the mythological summary of our ancestors' conscious intellectual experiences as they progressed from the collective unconscious womb of African root hominid consciousness itself. This is an epigenetic upward pull of the superconscious, the ascension of individual/group life force, following an earlier descent of light through space into matter and gravitational condensation of Black nanodiamond interstellar laden gas clouds into protostars. It represents the involution of spirit down into matter and the evolution of matter back into spirit in the great cycle of Divine Manifestation as seen in the sublime vision of Abydos.

From a cosmogenetic perspective, we witness the main sequence of star evolution: a star with a carbon core initial stage, flowering into a supernova and seeding interstellar gas clouds with carbon through a self-organization of complex carbon-based organic molecules in interstellar space, particularly hydrated-water-laden comet bodies coated with surface organic molecular matter seeding planetary surfaces with living organic dark matter. This self-organizing, biophysical carbon-rich molecular unfoldment moves progressively through evolution up through the Mineral Kingdom, Plant Kingdom, and Animal Kingdom to the human species and beyond. Thus, the Kemetic hill can be seen as a symbol of an erect individual human being as a progressive epigenetic unfoldment.

"AS ABOVE, SO BELOW": INNER BRAIN STRUCTURES AND OUTER SYMBOLIC FORMS

> Anu, 𓊖, i.e.,
>
> The cult of the standing stone, or pillar, was probably older than the cult of Rā, and the old name of Heliopolis is Anu, 𓊖, i.e.,

In Kemet, the oldest site of the study of the sun god was the city of Anu, later named Heliopolis by the Greeks and Romans and named On in the Christian Bible. According to Budge (Budge, 1923), it was at Anu that from time immemorial there existed a temple dedicated to the sun god, and this same temple was supported by an ancient college of priest/scientists who from a very remote period were renowned for their wisdom and learning.

Critically, Anu is also the name of the ancient Twa people of Africa, related to the early Homo erectus, who over 2 million years ago migrated from the Khuiland great lake regions to Ethiopia, Egypt, and into every continent. Thus, the seminal ideas and concepts of the African Kemetic priests of Anu are reflections of the germination of these ideas over several million years of the study of the sun, light, and the various forms of light to matter.

These priests called their sun god Tem or Atem, and the supremacy of Tem was defined in the various versions of the Book of the Coming Forth by Day (Book of the Dead), which reads in the 17th chapter, "I am Tem in his rising. I was the only one. I came into existence in Nenu [Nu, Nun, space]. I am Ra when he rose for the first time. I am the great god who created himself from Nenu, and who made his names to become the gods of his company. I am Tem, the dweller in his disk or Ra in his rising in the eastern horizon of the sky. I am yesterday; I know today. I am the Bennu [Black Phoenix] which is in Anu, and I keep the register of the things which are not yet in existence."

The ancient scientists at the old Kemetic university of Anu defined Tem as a man-god who absorbed the qualities of earlier African conceptions such as Heru-ur, the old sky god, sun by day and between sunset and sunrise, Heru-Khuti (Horus in the two horizons); Khepra, the sun during the hour that precedes the dawn; and Tem, the setting sun. Thus in the 17th chapter of the Book of the Coming Forth by Day, the earlier image of the sun god Tem showed the image of the sun god Ra. Budge noted that the importance of the priest scientists of Anu's concept of Ra was clearly shown in the third to fourth Dynasty by the use of the name of Ra of the Neso bat for the builder of the second (Khaf-Ra/Khephren) and third (Menkaru-Ra/Myercinus) pyramids at Gizah. It was this same father Tem god, later named Ra, earlier named Ptah, that presided over the company of other phases of god/light in the forms of Shu, Tefnuti, Geb, Nut, Osiris (Wosir), Isis, Set, and Nepthys.

The Kemetic pyramid text from the Kemetic Old Kingdom, earlier than 5,000 years ago, does record the ancient Anu African scientists' concepts related to melanin and neuromelanin in the concept of the black stone, or Ben-stone, ofrose granite. In the pyramid text, II, N.663, p.372, there is written how the spirit of the sun visited the temple of the sun from time to time in the form of a BlackBennu bird and "alighted on the Ben-stone in the house of the Bennu in Anu."

The Spirit of the Sun

visited the temple of the sun from time to time in the form of a Bennu bird, and alighted "on the Ben-stone,[1] in the house of the Bennu in Anu"; in later times the Bennu-bird, which

Pyramid Texts, II. N. 663, p. 372.

The Bennu bird was known by the Kemetic priests at Anu as the soul of Ra, or the Bennu Bird (Phoenix Bird). This was a symbolic model of the human Anu form, in which light entered the birdlike third ventricle, House of the Bennu, threshold cerebrospinal fluid brain system and alighted, thereby becoming activated there in the Ben-stone of neuromelanin, in the Black dot, locus coeruleus by the light activation and Black neuromelanin threshold translation of light along the entire 12 gate/12 black nuclei of the black neuromelanin amenta tract, the Kemetic I-33 tissue of Horus of the human brain C.S.F. ventricular system, brainstem, and spinal column.

Last, we so clearly see the common African origin of all humanity in our common Anu root ancestral viewing of the primordial images of Osiris, Khenti, Amenti, and the triune god of Osiris of the Osirian resurrection, including Seker, the old Death god of Memphis (Budge, 1923).

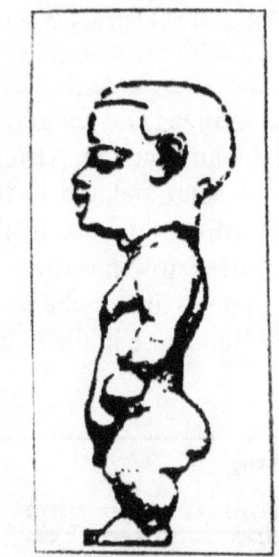

The triune god of the Osirian Resurrection. The three members of his triad were Seker, an old Death-god of Memphis; Ptah, a Creation-god of Memphis; and Osiris, the vivifier of the dead.

Osiris Khenti Amentt, god and judge of the dead and lord of the Other World.

Accordingly, one must also consider the meaning of the architectural symbolism of the Great Pyramid of Ghiza in Egypt. This is a structure that contains an uppermost all-black room of Black Rose Granite, the so-called King's Chamber that sits directly above the Queen's Chamber. For if the Great Pyramid is a symbolic model of the erect human form, it is comparable to the watery all-black room, third ventricle, located in the mid-brain limbic system structures of the human brain.

Another clue to the symbolic meaning of the Black Rose Granite room of the King's Chamber can be found in the architectural symbolism of the Black Rose Granite in the temple of Abydos. In this case, Black Rose Granite was used as the bottom, or threshold, stone of the four-sided gate/door entrance into the temple from the outside world into the outer courtyard. This same Black Rose Granite was also seen as the bottom, or threshold, stone in the four-sided gate/door between the outer courtyard and inner courtyard.

This Black Rose Granite is clearly seen functioning as a threshold, or symbolically trans formative gateway, over 5,000 years before the current era. Moreover, it was known that there was something black in the heads of humans that was not just a threshold for one line of movement but was black on all sides and a gateway to 360° expansion of human consciousness. A Black hypercube of unseen planes coming into vibratory manifestation, if you will. This has historic and dynamic allusions to awakening, to flight, to the mysteries of transformation and translation. The Islamic holy stone, or Kabaa, at Mecca is a black cube believed to have fallen from the stellar abyss with the capability of completely transforming the human spirit, which will burn its dross in the fires of meditation and purification.

Furthermore, upon close, visual inspection of Black Rose Granite, one can clearly observe a Black Stone with numerous streaks and zigzag lines of red, as if it were electricity, energy, or light being born out of blackness. This is the luminous Blackness alluded to before. This is the symbolic meaning of the Kemetic scientists who intentionally used such a stone clearly as a threshold entrance to the temple of Abydos, the site where the head of the perfect Black God of Amenta, Wosir, was buried in this Kemetic Holy Land. Was this the symbolic meaning behind the use of the same Black Rose Granite in the King's Chamber of the Great Pyramid?

Last, and of great relevance, come the inscriptions of the upper register of the right panel of the second shrine from the tomb of Pharaoh Tutankamen (Piankoff, 1977; Wimby-Jones, 1982).

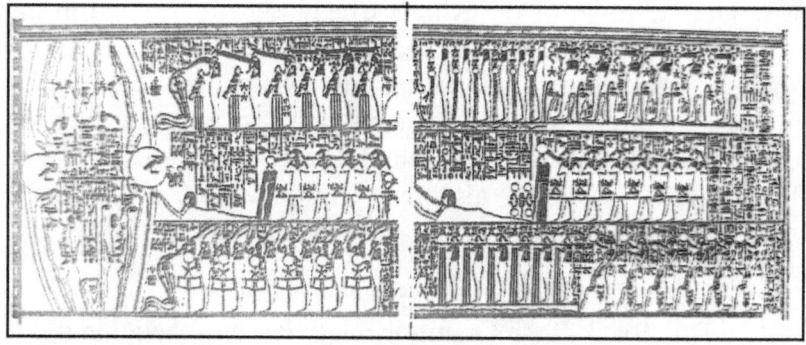

Upper Register-First group of divinities:
1. The morning (2); The Praiser (3); The Opener (4); The Keres (5); The Incomplete One (6); The Corruptible Flesh

Second group of divinities:
1. Head of Horus (2); Face of Horus (3); Neck of Horus (4); I-33 Tissues of Horus; (5) Inner Eye; (6) The Doorway

Third group of divinities:
1. Submerged One (2), (3); Ejaculator (4); Inundator (5); Babe in Swaddling Cloth (6); The Morning Bark of Ra

These are critical references to the luminous dark living melanin system present in the head that has come down intact to the present era from the pre-"white" supremacy time of the 18th Dynasty, 1349 B.C.

First, the upper registry, or first group of divinities, clearly pictures a star passing rays of light into the midforehead location of the pineal gland, the site of the light-sensitive inner eye, where the hormone serotonin during daylight, and during night, with star light/moonlight, the hormone melatonin is released. Its release has the effect of increasing the activity of melanin systems throughout the body.

The second group of divinities, fourth figure, named the Neck of Horus, may define the site of the human vocal cords, the site of the production of the spoken word, and especially the song word. This site of song production, of human music, defined the ancient study of the role of music and of harmonic resonance in elevating melanin systems and evoking the soul ascension through harmonic sound (Janata, 2002; Zatorre, 2002).

The second group of divinities, fourth figure, I-33 tissues of Horus, is a clear reference to the spinal cord with 33 vertebrae, which when stood erect, has a head and lateral eyes atop an erect spinal column and a bottom that stands level with the gonads, rectum, and genital organs, the womb of life.

The second group of divinities, fifth figure, is the inner eye, a clear reference to the pineal gland, an actual phylogenetic 3rd and 4th eye in the lower vertebrae, and the eye of inner vision in humans, the posterior floor of the third ventricle.

The second group of divinities, sixth figure, The Doorway, is again a reference to the mystery that in the head there is a doorway. Last, in the third group of divinities, the sixth figure, Morning Bark of Ra, may be a deeper aspect of the doorway, in144 that as the doorway ascends it becomes a vessel of vast travel, a literal star ship, the Morning Bark of Ra, a black star gate of inner vision.

Thus, following a review of the right panel, second register, from Pharaoh Tutankamen's tomb there is evidence to suggest that the Black Room, the King's Chamber of the Great Pyramid, was symbolic of a black chamber in the brain of humans that is a doorway, that is to say, a threshold entered into by the eye of inner vision that allows vast ascension; star travel; the Morning Bark of Ra, the perfect one, the goal of unity with the light (Blackshaw, 1999; Kume, 1999; Pickard, 1982; Moore, 1995).

Again, throughout the shrines of Tut-Ankh-Amon there are many references to Tut Ankh-Amon as the ruler of Heliopolis of the south. This clearly identifies the authors of the texts of the shrines as the great and very old college of priests/scientists of Heliopolis, or Anu of ancient ancestry and the ancient tradition of African scholarship in the study of psychology.

CURRENT AREAS OF SCIENTIFIC STUDY OF MELANIN AND NEUROMELANIN

The current era of scientific study has added much to our understanding of the meaning and function of both melanin and neuromelanin. These are the areas of physics, anatomy, embryology, endocrinology, and psychopharmacology. From physics, McGinness (McGinness, 1974) reports on both natural melanin, isolated from melanosomes from a human melanoma at autopsy, and synthetic melanin, produced by the enzymatic action of mushrooms' tyrosines.

In both cases, when exposed to an electrical current, melanin demonstrated the threshold-switching properties of an amorphous semiconductor. The electrical properties of all the melanin preparations were essentially the same. Threshold switching refers to the ability of Black melanin to switch from an "off" state of low conduction to an "on" state of high of electrical conduction (semiconductor). Furthermore, McGinness, in 2001 Post Publications, noted "... another missed opportunity-melanins give a flash of light when they switch-dearly electroluminescence, though we did not completely understand its significance at the time."

Again, one can more than marvel at the ancient African use of the Black Rose Granite in the threshold bottom of the Gate/ Door entrance to the temple of Abydos. Could the Black Rose Granite, which contains flecks of red, have been placed in the threshold to symbolize the essential function of melanin as a threshold that under certain energetic conditions switches to high levels of energy and information conduction and while doing so, simultaneously radiates light into other dimensions (electroluminescence)?

Water hydration was reported to play a critical role in the ability of melanin samples to undergo threshold switching. If the samples were dried for 30 minutes at 200° C., they would not switch until re-hydrated and dried at room temperature. Water was found to lower the activation energy of conduction by altering the local electrical constant of the material. The conductivity of the dopa melanin and isolated melanosomes was found to be high, 10.5 (ohm cm), with a resistance of 104 ohms for a sample 1 mm thick. The conductivity in the "on" state was increased by a factor of 100 to 1,000. Melanins were found to switch at 3.5 x IG2 volt/ cm and through at least 1 cm of material.

As a result of these findings, McGinness reported that the threshold semiconductor role of melanin was evident in the appearance of melanin in living organisms at locations where energy conversion, or charge transfer, occurs (the skin, retina of the eye, mid-brain structures of the limbic system, and the inner ear).

From a neurophysiological perspective, one can appreciate the physiological function of neuromelanin. Wherever it is found in its multiple anatomical sites, it serves in the role of energy threshold, that is to say, of actual "energy conversion," or threshold change transfer.

In the lateral eyes, the same threshold role of melanin is seen at the site of retinal-pigmented epithelium (RPE). One form of charge transfer is the conversion of rod or cone vision pigments that entrap external world photons into an electrical charge for passage into the optic nerve. Another threshold role is the transfer of the rod/cone visual pigment carrier of the photons into an RPE peptide signal, melanoprotein signal, and a neuroendocrine signal. The melanoprotein signal then is carried through the blood to a mid-brain site, the suprachiasmatic nucleus, SCN, in the wall of the third ventricle.

This melanoprotein signal triggers the SCN to pour forth through the L.C., the 12 neuromelanated nuclei of the Amenta nerve tract I-33 Tissue of Horus, and it pours forth a vast waterfall-like cascade of hormonal effects during the 24-hour cycle of circadian rhythm. This 24-hour cycle has a 12-hour rise to a high level of these hormones followed by a 12-hour decline to a low level of hormones produced by most of the endocrine glands and other organ systems of the entire body. It is this environmental photon/ RPE melanoprotein/ SCN/LC-Amenta trance circadian rhythm that is believed to be so clearly illustrated by the fact that melanin is a threshold door that allows the microcosmic individual human being to be in rhythm with the external macrocosmic world light/dark cycles of environmental lighting.

Many western scientists of the current era now define Neuromelanin (NM) as a Complex polymer pigment found in the catecholamine neurons of the human brain in the locus coeruleus (LC), substantia nigra (SN), but seldom discuss the 10 pigmented neuron centers in the brain stem. Accordingly, they identify two major forms of NM as being the result of the spontaneous auto oxidation of dopamine as aminochromes, dopaminochromes in the SN and adrenochrome in the LC. Whereas the SN NM has been reported to be a copolymer of black eumelanin and pheomelanin produced from oxidized cystinly-DOPA products. In addition, in these brain sites the NM polymer pigment includes benzothiazine heterocyclic products, glycidic lipid matrix, complex protein compounds, and inorganic metals such as iron (Double, 2000).

Modern scientist have now discovered a specific role for NM in the illness of Parkinson's disease (PD) that actually illustrates a general role of NM in cellular physiology. There are reports that in the illness PD there is a loss of NM neurons in the SN, iron bond to NM granules, a 70% reduction in SN NM neurons, and a 69% increase in the free radical iron in the SN in PD patients. The increase levels of free radical iron in the NM neurons of the SN is now thought to reflect an over loading of NM with free radical iron resulting in NM neuron damage in the SN.

Moreover, the discovery of the relationship of NM in the SN to iron is now believed by several modern scientists to reveal a general function of NM as one cellular detoxification. This may be better defined as an actual recycling of potentially dangerous processes such as free radical iron in neurons, free radical scavenging iron induces peroxidation, and conversion of potentially toxic dopaminergic metabolic products (Faucheux, 2003). Again, melanin and NM in its multiple anatomic sites serves a role of energy threshold, or energy conversion, the recycling of energy threshold charge transfers. Electroluminesce, NM translates multiple chemical events into light.

NEUROMELANIN IN EMBRYOGENESIS: THE UNFOLDMENT OF THE ORGAN SYSTEMS

When we return to the enigma of the all-black chamber in the inner brain structures, we should remember, as outlined in Chapter 1 by Professor T. Owens Moore, our neuroembryology and our neuroanatomy as we developed in the womb. First, the biological history of the neuromelanin all-black chambers of the brain begin before conception. Melanin is present in the tail of the father's spermatozoa. Melanin is present in the eggs of the mother's ovaries. The egg, following release from the ovary, is fertilized, united with the sperm in the mother's fallopian tube. The fertilized ovum becomes a rapidly multiplying ball of cells that within hours develops into a morula, or blackberry, so named because it looks like a blackberry.

The outer layer, the cells of the ectoderm, is black. The future epidermal cells maintain their critical relationship to melanin as a site for threshold of energy conversion and/or threshold of charge transfer. The ectoderm undergoes vast migrations into other anatomical sites during this embryological stage of development of the human form from its early black, melanin covered seed of life.

It is critical to note that during the first 28 hours following conception there occurs in the morula an invagination of the black dot of the morula's black ectoderm surface inward to form a tube, the neural tube. The black dot tip of this inward moving neural tube balloons out to become the brain. The tube itself becomes the spinal column. On the third day following conception, the morula moves from the fallopian tube into the uterus and bonds to the wall of the uterus. It is crucial to note that the first two hormones produced on this 3rd day are MSH, melanocyte stimulating hormone, and HCG, human chorionicgonadotrophin hormone. Along the line of the neural tube there is an outer layer surrounding the neural tube, the neural crest. The neural crest is a site of many blast cells, those "immortal archetypal daughter cells" from which entire lines of other cells originate, such as the haemopoeitic cells, white and red blood cells, gonads (spermatozoa, ova), bone cells, and especially endocrine glands and exocrine glands that line the GI tract, urogenital tract, and respiratory tract. These vast arrays of endocrine and exocrine glands are grouped together in the broad category of the APUD cell series by virtue of the fact that they all originate from the neural crest. Moreover, they still display a marking of this vestigial origin from black ectoderm by virtue of possessing the same decarboxylase enzyme that allows a process charge transfer of the carbon atom from an early carbon-atom-rich melanin ectoderm. The list of endocrine glands includes the hypothalamus, pineal, pituitary, mast cells, thyroid, parathyroid, thyrocalcitonin cells, pancreas, adrenals, and the gonads (Pearse, 1969; 1976).

Whereas the neural tube itself in the brain becomes the C.S.F. (cerebral spinal fluid), the ventricular system that extends into the spinal cord as the central canal is lined by the C.S.F. contracting neurons (Vigh, 1975, 1977, 1980). In the brainstem of humans, these 12 pigmented neurons, of which there is a phylogenetic range of an increasing number of pigmented nuclei, only humans have all 12 pigmented, the locus coeruleus being unique in humans. Thus, there is increasing pigment upward along the neural tube. This is the I-33 Tissue of Horus, the Amenta nerve track with 12 sites of neuromelanin. The Amenta nerve track forms a neuromelanin covering of the third ventricle.

NEUROGENESIS

Neurogenesis is the creation of new neurons that are added to specific regions of the brain's association cortex. Neurons are the cells of the nerve tissue, with extensions running to and from them carrying and communicating information in the form of electrical impulses about the operation of the body's organ systems, especially the brain. Axons are those nerve fibers that conduct these impulses away from the body of the nerve cell, whereas dendrites are those nerve fibers that transmit electrical impulses toward the nerve cell body.

The vast universe of neurons and their innumerable interconnections comprise the mystery of the brain. New neurons that are added in neurogenesis are involved in learning and memory (Gould, 1997, 1999). The new neurons originate in the granular layer of the dentate gyrus from the wall of the lateral ventricles sub-ventricular zone (the melanin-lined old morula/gastrula ectoderm of the lateral ventricles) and then migrate downward along radial glial fibers to the L.C.-modulated hippocampus and amygdala and then migrate upward as L.C.-pigmented neurogenesis neurons to seed the cortical association areas of the prefrontal, postfrontal, inferior temporal, and parietal cortex. This course of neurogenesis occurs over the course of 21 days. Neurogenesis is believed to be significantly regulated by psychosocial stressors (Gould, 1997). Throughout life in the brains of humans and other mammals the two areas of the brain that generate new neurons are the olfactory bulb (Ng, 2005) and the dentate gyrus of the hippocampus (Gage, 2000). In both cases new neurons are generated from the subventricular zone of the lateral ventricles.

Within all humans, regardless of their surface racial or ethnic identification, there exists the same brainstem that contains the same black neuromelanin nerve tract, the Amenta nerve tract with black neuromelanin. These 12 centers are the (1) locus coeruleus, (2) substantia nigra, (3) brachialis, (4) paranigralis, (5) intracapsularis, (6) nervi trigemini, (7) mesencephalius, (8) pontis centralis orates, (9) tegmenti pedunculopontis, (10) parabrachialis, (11) medialis dorsomotor, and (12) the retroambigualis (Olszewki, 1964; Marsden, 1961; Bazelon, 1968; Feinchel, 1968; Forrest, 1972, 1975; Lacy, 1981, 1984; Sandyk, 1991; Santamarina, 1958; Vigh, 1975, 1977, 1980; McGinnes, 1989; Lindquist, 1987; King, 1994). All animals with spinal columns, the vertebrates, have varying degrees of neuromelanin pigmentation of these 12 centers. The earlier life forms such as fish, amphibians, and reptiles have fewer of the 12 centers pigmented. The phylogenetically advanced forms have more of the centers pigmented, and mammals have the largest number of neuromelanin-pigmented brain melanin centers. Within the primate family, the near human chimpanzee has 11 of the 12 centers containing the deepest black neuromelanin pigmentation.

Humans are the only primate, and only vertebrate life form, to have deep neuromelanin pigmentation of the twelfth brain center. This is the locus coeruleus. The locus (Sanskrit/point) coeruleus (Latin/black) means black dot. This 12-step brainstem tract of neuromelanin is named the Amenta nerve tract. It is a neuromelanin nerve tract that in humans is found in the center of the midline of the brain's spinal column that surrounds the C.S.F. (cerebral spinal fluid) ventricular system and by virtue of its morula ectoderm origin is found as a sub-ventricular layer below the epithelium/C.S.F. containing the neuron lining of the ventricular system. It runs from the wing-like lateral ventricle, especially surrounding the mid-line 3rd ventricle aqueduct of sylvius, and 4th ventricle, leading into the central canal that runs through the entire spinal column.

Significantly, as vertebrate life forms developed increased brain complexity, there was progressively increasing pigmentation of centers more anterior and higher up the C.S.F. ventricle tree. It is noted that as humans evolved from Australopithecus to Homo erectus to Homo sapiens sapiens, our own species, the expansion of the cerebral cortex has witnessed increased pigmentation of the L.C. and an expansion of the cerebral cortex and the neurogenesis of pigmented cortical neurons.

Cells of the locus coeruleus (L.C.) provide the principal noradrenergic, norepinephrine nerve supply to many areas of the brain, especially the cerebral cortex, hippocampus, cingulate gyrus, and amygdala areas, which make up the major portion of the hippocampal limbic cortex (Amaral, 1977; Kobayshi, 1975). Thus, in the neurogenesis migration of new neurons from the lateral ventricle through the limbic system, it is certain that such neurons are profoundly encoded by input from the Amenta nerve track/ black neuromelanin nerve tract through its uppermost L.C. center.

Neuromelanin charge transfer of discrete information signals is conducted along the neuromelanin Amenta nerve track into the migrating new neurons. This is the emotional coloring or tagging of thought, the development of much higher order neural melanin "chips" if you will. The hard-wired new neurons are then seeded into four critical cortical association sites to expand and further uplift the thought patterns of the individual and group in its collective conscious development. In this manner, emotions at times, especially at sacred or "awe inspiring" times, can be signals that may expand consciousness to higher levels of expression. Similarly, the L.C. supplies part of the norepinephrine found in other brain areas, such as the hypothalamus, thalamus, pineal glands, habenula (deep pineal), cerebellum, lower brainstem, and the spinal cord.

However, stress can abort neurogenesis levels, leading to increased cell death of these new neurons before such neurons complete the cycle of development and form a lateral ventricle birth through limbic system modulation, followed by migration to cortical association areas. During such stress events, there is increased norepinephrine and uncoupling of serotonin from norepinephrine.

In the normal person, with a normal low level of norepinephrine, increased serotonin and norepinephrine, and lower levels of serotonin, psychosocial stress inhibits new neuron cell proliferation. Furthermore, Gould reported "rapid suppression of cell proliferation by a threatening experience, conveyed by cues from different sensory modalities (visual), is a correlation of the dentate gyrus that is common to mammalian species that undergo adult neurogenesis" (Gould, 1997).

Thus, on the issue of threshold switching, there is an even larger array in the role of melanin functions as reported by Breatwatch (Breatwatch, 1988). These are the roles other than the skin or eye (Ptah, 1978; Drager, 1986) and the site of the inner ear (Meyer Zurn Gottesberge, 1988). These are the neuromelanin I-33 tissue of melanin sites in the 12-pigmented nuclei of the Amenta nerve tract where neuromelanin functions in the multiple roles of threshold switch.

Neuromelanin appears to provide a threshold for electrical signal conversion redox capacity, an electron transfer agent, an amorphous semiconductor threshold switch, and an electron-photon couple. It also acts as an accumulator of drugs and metal ions and has carbon exchange properties, a reservoir for trace elements, and a sink for free species. It also selectively refines energy signals from one state to another within the living neural system.

A consideration of the ancient African symbolic metaphor of the Black Rose Granite cube-shaped King's Chamber of the Great Pyramid, with an open empty rose granite sarcophagus, leads us to the black inner brain chamber, with each corner standing as a symbolic reference to the neuromelanin, locus coeruleus, Amenta nerve area, and third ventricle. This dark inner brain chamber of all humans is indeed an enigma, a supreme mystery of Blackness. Please consider this meager theory as prayer or mediation, as a reaching up and trying to understand such visions of Blackness.

To begin, in the third ventricle there is a very critical linkage of the C.S.F. fluid of that ventricle, which contains high concentration levels of hormone signals from the various endocrine glands (the pineal, pituitary, gonads, urogenital gland) to the GI tract, respiratory tree, etc. Secondly, there is a likely translation and further modulation of such signals by the C.S.F. contracting neurons that are then conveyed by the L.C. into higher thought patterns, as indicated by new neurons in the association cortex.

Further selective refinements of energy signals of light and gaseous liquid, lead to higher emotional, sexual, and auditory experiences that further elevate thought patterns and continue to physically charge neurons so that this high order sensing and unfoldment leads to inner vision, ascensions of the five senses, and eventually unity with light.

This is all to suggest that a dormant field singularity may exist in the third ventricle, which, at a critical threshold, allows the inner vision to undergo a 360° expansion. This Bark of Ra, in the classical Kemetic tradition of meditation, allows for the emergence of advanced sensory and intuitive experiences.

CLINICAL CORRELATIONS OF I-33 TISSUE OF HORUS/AMENTA NERVE TRACT/ NEUROMELANIN

When considering the underlying conceptual and clinical aspects of this brainstem neuromelanin operation in all humans, Naim Akbar (1985) suggested that the European branch of this African root species has undergone a subtle kind of de-evolution that has 2 basic tendencies: (1) sexism, or the patriarchal fear and denigration of the feminine; and (2) racism, the fear and denigration of bloodline African peoples. In this context, it is important to note that with many European-Africans there is indeed, clinically speaking, a legacy of pineal gland calcification, which has resulted in lower levels of pineal melatonin and pineal serotonin (King, 1994). It is reported to be about half of the levels found in African populations (King, 1994b; Pelham, 1973; Vaughn, 1976).

Thus, the critical differences between high skin-level melanin (African-African) of the parent population and their children and European-Africans of low skin melanin, have resulted in a shift in the population to a lower level of melatonin. This critical finding of low pineal melatonin and low pineal serotonin in a European-African dominated world may go a long way in explaining the tendency of White supremacy programs that seek and continue to enforce designs that perpetuate low pineal melatonin/serotonin in themselves and their dominated African-African parents.

A condition or stage of high fear of inner vision born of an aborted inherent identification with the psychological Blackness results in mental slavery. In a world order dominated by the paradigm of White supremacy, the White mental slave attempts to drive the fear of the psychological Blackness conversation with one's bloodline genetic Black ancestors (and the same mental slavery) into others. Accordingly, in the pre-White supremacy world of Pharaoh Tutankamen, the study of the operation of the I-33 Tissue of Horus, the Amenta nerve track, was the development of inner vision through the three grades of students with high melatonin/ serotonin and conscious neurogenetic propagation and learning even in the face of fear. Whereas in the present White supremacy dominated world there is high norepinephrine (NE), low pineal melatonin/serotonin propagation, and aborted neurogenesis. It is this incomplete neurogenesis in adulthood that stops learning and keeps us worshiping the false images of the Divine in the face of fear.

There is a fundamental effort embedded in the paradigm of White supremacy to enforce fear and promote images of immense pain, misery, and castration among Black people, all replete with a real history of lynching of people of color, all the while attempting to hide the higher sciences that are born of this inner Blackness. The edifice of a fragmented material science is then used as propaganda in a dismal attempt to obscure the educational programs that enhance development of this inner vision.

Yet, in the treatment of people in the fear-based Eurocentric molds that have high NE, low serotonin the technique remains the same. The treatment is to review the individual and family history of the client to identify those symptomatic and emotionally laden trigger events and sensory experiences that are historically linked to the fear-based trauma, the conditions present at the time of the trauma, and the inappropriate guilt or blame assumed by the person (which results in a damaged Eye of Heru).

Then comes the review of that person's life and support system in order to find the pieces of the inspired Eye and to knit the damaged lateral Eye of outer vision into the awakened Eye of inner vision. A dual cultural hemispheric multisensory experience of intense, logical, and Kemetic "Way of the Heart" focuses paths that blend together by finding the Black Dot pupils, those core creative passions given to one by the God force (Higher Power), and because of their affinity with light itself, possessed by neuromelanin-modulated neurons. This is the literal Black Ben Stone, which is daily light infused by the Bennu Bird, light itself, the soul of Ra (Higher Powers). This has been from time immemorial the human family's epic struggle: to locate other awakened humans, symbolized by the stories and myths of angels, genies, etc., and to follow their disciplines and realize the promise that some day in each one of us this power for transformation will be awakened.

Upon review of the right panel of the second shrine, it appears that the upper register details a process of epigenetic enfoldment of the three different grades of inner vision. During the development of inner vision, all three planes in the same person exist at the same time but in different dimensions; yet they concurrently interpenetrate each other at the same moment. On the neophyte plane there is a struggle to hold onto the flesh, a belief that the material world is real, and there results a fear of the loss of the physical body. It is this mental fear that dreads the decay of the flesh, the physical self of creation and instrument of embodiment.

In the next, or high stage, the stage of Intelligence, there is the development of knowing that the physical body is but a doorway through which the soul interacts with of Eye of Heru and the Eye of illumination. The inner vision is awakened and moves like a river through the soul and has communication with and experiences the existence of many other worlds. Through this doorway is housed, by the Morning Bark of Ra, the awakened spiritual consciousness, the Eye of Heru and our capacity for travel and communication with the celestial family of ancestors and spirits in the higher realms of mind and consciousness.

In summary, the operation of I-33 Tissue of Horus, the neuromelanin nerve tract, makes known this ancient African education system (James, 1954; King, 1990; Budge, 1967; Welsing, 1990; Nobles, 1976; Faulkner, 1969; Piankoff, 1977; Clark, 1975; McGee, 1976; Moore, 2002; Bynum, 1999) that promotes the progressive refinement of passionate emotional processes and results in increased melatonin/serotonin and neurogenesis of high, new neuronal development even in the face of fear.

The seeding of such higher-order modulated emotion means that the ascension cortex literally transforms the brain/ whole body to allow an ascension of not only vision but all of the soul's capacity for unity with light and communication with what has been known since ancient times as the immortals (ancestors). This is the genius level of ideal perfection, the primordial realm of archetypal Intelligences envisioned on the banks of the Nile millennia ago. This is the union of oneself with the supreme God, traveling in the Morning Bark of Ra.

Finally, the "frenzy of shouting when the spirit of the Lord passed by, seizing the devotee and making him mad with supernatural joy (Holy Ghost), was the essential of Negro religion and the one believed more devoutly believed in than all the rest" (Dubois, 1906). This" supernatural joy" is to travel in the many realms of the living, knowing well the full numerical symbolism and associated multilevel sensory experiences of the entire line of humanity's ancestors in light, song, and rhythmic harmony. For it is through our melanin with all its properties that African peoples and, perhaps by extension, all peoples-since the African genotype is the template for our species-experience that supernatural joy, those high-feeling states, and come directly to experience being "touched by the spirit." It is that fleeting glimpse of our souls that is the inner vision of the genius Ka, a genius and light that now reawakens (Gregory, 2003).

REFERENCES

Adams, H. H., 1994. MA'AT: Returning to Virtue-Returning to Self. Chicago.

Akbar, N., 1985. "Nile Valley Origins of the Science of the Mind. Nile Valley Civilizations," in Proceedings of the Nile Valley Conference, 2, 120-132. Atlanta: Journal of African Civilizations.

Allen, T. G., 1974. The Book of the Dead or Going Forth by Day Chicago: University of Chicago Press.

Anders, E., and Zinner, E., 1993. Metoritics, 28,490.

Andrews, S. M., 1989. Color Me Right ...Then Frame Me in Motion. TN: Seymore-Smith.

Ani, M., 1994. Yurugu: An African-Centered Critique of European Thought and Behavior. Trenton, NJ: African World Press.

Assagioli, R., 1965. Psychosynthesis: A Manual of Principles and Techniques, 1-20. New York: Viking Press.

Astraw, B., White, T., et al., 1999. "Australopithecus garhi: A new Species of Early Hominid from Ethiopia," Science, 284, 629-635.

Barnes, C., 1988. Melanin: The Chemical Key to Black Greatness, vol. 1. Houston: C. B. Publishers.

-1993. Jazzy Melanin: A Novel. Houston: Melanin Technologies.

Barr, F. E., 1982. Melanin and the Mind-Body Problem. Berkeley, CA: Institute for the Study of Consciousness.

-1983. "Melanin: The Organizing Molecule." Medical Hypothesis 11, 1-140.

Bazelon, M., Feinchel, G. M., 1968. "Studies on Neuromelanin 1: A Melanin System in the Human Adult Brainstem," Neurology, 18, 817-820.

Ben Jochannan, Y., 1972. Black Man of the Nile and His Family. New York: Alkebu-Lan Books Associates.

Bernal, M., 1991. "Black Athena: The Afroasiatic Roots of Classical Civilization," in vol. 1, The Archaeological and Documentary Evidence. New Brunswick, NJ: Rutgers University Press.

Blackshaw, S., and Snyder, S. H., 1999. "Encephalopsin: A Novel Mammalian Extraretinal Opsin Discretely Localized in the Brain," Journal of Neuroscience, 19, 3681-3690.

Bradley, J.P., and Brownlee, D. E., 2002. "Cometary Particles: Thin Sectioning and Electron Beam Analysis," Science, 231, 1542-1544.

Breatwatch, A. S., 1988. Extra-cutaneous Melanin, Pigment Cell Research, 238-249.

Brunet, M., Guy, F., et al., 2002. "A New Hominid from the Upper Miocene of Chad, Central Africa," Nature, 418, 145-151.

Budge, E. A. W., 1969. The God of The Egyptians, vols. 1 and 2. New York: Dover Publications.

-1923. Tutankamen, Amenism, Atenism and Egyptian Monotheism. New York: Dell Publishing Company.

-1967. The Book of the Dead. New York: Dover Publications.
Bynum, E. B., 1984. The Family Unconscious: An Invisible Bond. Wheaton, IL: Theosophical Publishing House.

-1999. The African Unconscious: Roots of Ancient Mysticism and Modern Psychology. New York: Columbia University Teachers College Press.

-1994. Transcending Psychoneurotic Disturbances: New Approaches in Psychospirituality and Personality Development. New York: Haworth Press.

Calif, N. M.A. "Past Ancestors, My Immediate Family, My Extended Family and My Future Ancestors."

Clark, X. C., McGee, P. P., Nobles, W., 1975. "Voodoo or I.Q.: An Introduction to African Psychology," Journal of Black Psychology, I, 2, 9-19.

Chruchward,A., 1913, 1986. The Signs and Symbols of Primordial Man. Evolution of Religious Doctrines from the Eschatology of the Ancient Egyptians, 2nd ed. London: George Allen & Co., New York: E. P. Dutton & Co.

- 1921. The Origin and Evolution of the Human Race. London: Allen and Unwin.

-1924. The Origin and Evolution of Religion. New York: E.P. Dutton & Co.

Clarke, J. H., 1999. My Life In Search of Africa. Chicago: Third World Press.

Cotzias, G. C., Papa Vasiliou, P. S., Van Woort, M. H., Sakamoto, A., 1964. Fed. Proc. 23, 713.

Diop, C. A., 1981. "Origin of the Ancient Egyptians," in General History of Africa, vol. 11 of "Ancient Civilizations of Africa,"

G. Mokhtar, ed. Berkeley, CA: University of California Press.

--1991. Civilization or Barbarism. West Port, CN: Lawrence Hill&Co.

Double, K. L., Zecca, L., Costi, P., Mauer, M., Griesinger, C., Ito, S., Ben-Shachar, D., Bringman, G., Fariello, R.G., Riederer, P., and Gerlach, M., 2000, "Structural Characteristics of Human Substantia Nigra Neuromelanin and Synthetic Dopamine Melanins" Journal of Neurochemistry, 75, 6, 2583-2589.

Drager, U. C., 1986. "Albinism and Visual Pathways," The New England Journal of Medicine, 314, 25, 636.

Dubois, W. E. B., 1906. The Souls of Black Folk. New York: Bantam Books.

Falk, D., Redmond, J. G., et al., 1999. "Early Hominid Brain Evolution: A New Look At Old Endocrasts," Journal of Human Evolution, 38, 695-717.

Faulkner, R. 0., 1969. The Ancient Egyptian Pyramid Texts. Oak Park, IL: Aris and Phillips, Bolchazy-Carducci.

-1978. The Egyptian Coffin Texts, 1-111. England: Westminster.

Feaucheux, B.A., Martin, M.E., Beaumont, C., Hauw, J.J., Agid, Y., and Hirsh, E.C., 2003, "Neuromelanin associated Redox-Active Iron is Increased in the Substantia Nigra of Patients with Parkinson's disease", Journal of Neurochemistry, 86, 1142-1148.

Feinchel, G. M., 1968. "Studies onNeuromelanin2: Melanin in the Brainstem of Infants and Children," Neurology, 18, 817-820.

Forrest, F. M., 1975. "The Evolutionary Role of Neuromelanin," West Pharmacology, 18,205.

Gage, F. H., 2000, "Memmalian Neural Stem Cells" 287, 1433.

Gould, E., 1999. "Serotonin and Hippocampal Neurogenesis," Neuropsychopharmacology, 21, 463-515.

Gould, E., McEwin, B. S., Tanapat, P., Glen, L.A. M., Fushs, E.., 1997. "Neurogenesis in the Dentate Gyrus of the Adult Tree Shrew .....," Journal of Neuroscience, 17, 2492-2498.

Gould, E., Reeves, A. J., Grazians, L.A., Gross, G. G., 1999. "Neurogenesis in the Neocortex of Adult Primates," Science, 286, 548-552.

Gregory, D., 2003. "The Legacy of Dr. Martin Luther King, Some thing Greater Than Fear or Hatred," Black Psychiatrists of

America, section V of the National Medical Association, Beverly Hills, CA.

Guderian, H., Panzer Leader, P. D. A., 1952. Cambridge, MA: Kopo Press.

Hill, H. G. M., Jones, K. P., and d'Hendecourt, 1998. "Diamonds in Carbon-Rich Proto-Planetary Nebulae," letter to the editor, Astronomy and Astrophysics, L41-L44.

Howe, S., 1998. Afrocentrism: Mythical Pasts and Imagined Homes. New York: Verso.

Jablonski, N. G., Chaplin, G. T., 2000. "The Evolution of Human Coloration," Journal of Human Evolution, 39, 57-106.

Jackson, J. G., 1970. Introduction to African Civilizations. New York: University Books.

James, G. G. M., 1954. Stolen Legacy. New York: Philosophical Library.

Janata, P., Birk, J. L., Van Hom, J.P., Leman, M., Tillmannn, B., Bharucha, J. J., 2002. "The Cortical Topography of Tonal Structures Underlying Western Music," Science, 298, 2167-2270.

Jantsch, E., 1980. The Self-Organizing Universe. New York: Pergamon Press.

Jorgenson, VG., 1988. Nature, 322, 702.

Kappelman, J., Swisher, C. C., et al., 1996. "Age of Australopithecus Africans from Fejej, Ethiopia," J. of Human Evolution, 30, 139-146.

Kershaw, I., 1998. Hitler 1889-1936. Hubris, NY: W.W. Norton.

-1999. Hitler, 1937-1945. Nemesis, NY: W.W. Norton .

Kimbel, W. H., Johnson, D. C., et al., 1994. "The First Skull and Other New Discoveries of Australopithecus africans at Hada, Ethiopia," Nature, 308.

King, R. D., 1990. The African Origin of Biological Psychiatry. Chicago: Lushena Books.

-1992. Kemetic Images of Light, Sunlight and Moon Light. Durham, NC.

--1993."Black Symbolism of the Unconscious: Part 1," review of Black symbolism in the collected works, vol. 20, C. G. Jung, 1933.

-1994. "The Pineal Gland, Melanin and Calcium: Pineal Gland Calcification in African-Americans, a Review of 1,622 Cases," scientific essay. Durham, NC.

-1994. Melanin: The Key to Freedom. Chicago: Lushena Books.

Kobayashi, R., 1975. "Biochemical Mapping of the Noradrenergic Projection from the Locus Coeruleus," Neurology.

Kume, K., Siram, S., Shearman, L. P., Weaver, D., R., Jin, X., Maywood, E. S., Hastings, M. H., and Reppert, S. M., 1999. "MCRY and MCRYZ Are Essential Components of the Negative Limb of Clock Feedback Loop," Cell, 193-205.

Lacy, M., 1981. "Neuromelanin: A Hypothetical Component of Bioelectric Mechanisms in Brain Function," Physiol. Chem. & Physics, 13, 319-324.

-1984. "Phonon-electron Coupling As a Possible Transducing Mechanism in Bioelectronic Processes Involving Neuromelanin," Journal of Theoretical Biology, 111, 201-204.

Leakey, G., 1995. "New Four-Million-Year-Old Hominid Species from Ranapoi and Allia Bay, Kenya," Nature, 376, 565-571.

Leigh, S. H., 1992. "Cranial Capacity Evolution in Homo erectus and Early Homo sapiens," American Journal of Physical Anthropology, 87, 1-13.

Lewis, R. S., Anders, E., and Draine, B. T., 1989. Nature, 339, 117.

Lindquest, N. G., 1987. "Neuromelanin and Its Possible Protective and Destructive Properties," Pigment Cell Research, 133-136.

-1973, Acta Radiology, 325.

Liou, J.C., Zook, H. A., and Dermott, S. F., 1996. "Kuiper Belt Dust Grains As a Source of Interplanetary Dust Particles," Icarus, 124, 429-440.

Luu, J. X., 1993. "Spectral Diversity Among the Nuclei of Comets," Icarus, 104.

Marsden, C. D., 1961. "Pigmentation in the Nucleus Substantia Nigra in Primates," Journal of Comparative Anatomy.

McGee, P., 1976. "An Introduction to African Science: Melanin, the Physiological Basis for Psychological Oneness," in L. M. King, et al., eds., African Philosophy: Assumptions and Paradigms for Research on Black Persons. Los Angeles: Fanon Center Publications.

McGinness, J., Cony, P., Proctor, P., 1974. "Amorphous Semiconductor Switching in Melanins," Science, 853-855.

McGinness, J.E.,1985. "A New View of Pigmented Neurons," Journal of Theoretical Biology, 115, 475-476.

McHenry, H., Berger, M., Lee, R., 1998. "Body Proportions in Australopithecus," Journal of Human Evolution, 35, 1-22.

Moore, R.Y, Speh, J. C., and Card, S. P., 1995. "The Retinohypothalmic Tract Originates from a Distinct Subset of Retinal Ganglion Cells," Journal of Comparative Neurology, 352, 351-366.

Moore, T. 0., 1995. The Science of Melanin, Dispelling the Myths Silver Spring, M.D.: Venture Books, Beckham House Publishers.

-2002. Dark Matters, Dark Secrets. Redan, GA: Zamani Press.

Ng, K.L., Li, J.D., Cheng, M.Y., Leslie, F.M., Lee, A.G., and Zhou, Q.Y., 2005, "Dependence of Olfactory Bulb Neurogenesis on Prokineticin 2 Signaling", Science, 308, 5730, 1923-1927.

Nicolas, B. R. J., Nicolas, R. A., 1998. "Biological Garbage or Jewels?" Scientific communication presented at the meeting of the European Society for Pigment Cell Research, Sept. 23-26, 1998, Prague. Pigment Cell Research, 11,233.

-1997. Speculating on the Band Colors in Nature, vol. XLV, 365. Atti dell Accademia. pontananiana, Giannini, Napoli.

Nicolas, R. A., 1997. "Colored Organic Semiconductors: Melanins Rend," Acc. Sc. Fis. Mat. Napolis, vol. LXIV, 325-360.

Nobles, W., 1986. African Psychology: Toward Its Reclamation, Reascension and Revitalization. Oakland, CA: A Black Family Institute Publication.

Nobles, W., 1976. "African Science and Black Research, The Consciousness of Self," in L. King, V Dixon, and W. Nobles, eds., African Philosophy: Assumptions and Paradigms for Research on Black People. Los Angeles: Fanon Research and Development Center Publications.

Nuth, J. A., and Allen, J. D. E., 1992. Astrophysics and Space Science, 196, 117.

Olswezski, J., 1964. Cytoarchitecture of the Human Brain Stem. New York: Sternnam Birjelow.

Pandey, S., Blanks, J. C., Spee, C., Jiang, M., and Fong, H. F. W., 1994. "Cytoplasmic Retinal Localization of an Evolutionary Homolog of the Visual Pigments," Experimental Eye Research, 58, 605-613.

Path, M. 0., 1978. "Phagocytes of Light and Dark Adapted Rod Outer Segments by Cultural Epithelium," Science, 203, 526.

Pearse, A. G. E., 1969. "The Cytochemical Ultra Structure of Cells ,"Journal of Histochemistry and Cytochemistry, 17, 303-313.

-1976. "Neuroendocrine Embryology and the APUD Concept," Clinical Endocrinology, S. Supplement, 2335.

Pelham, W., Vaughn, G., Sandock, K., Vaughn, M., 1973. "Twenty four Hour Cycle of a Melatonin-like Substance in the Plasma of Human Males," Journal of Clinical Endocrinology Metalo., 37, 341-344.

Petrie, W., 1939. The Making of Egypt, fig. 1, 86-89. London.

Piankoff, A., 1954. The Tomb of Ramses VI, 208, "Bollingen" series, XL-I. New York: Pantheon Books.

-1977. The Shrines ofTutankamen, "Bollingen" series, XL. Princeton, NJ: Princeton University Press.

Pickard, G. E., 1982. "The Afferent Connections of the Suprachiasmatic Nucleus ," Journal of Comparative Neurology, 211, 65-83.

Proctor, P., 1972. Physical Chem. Phys., 4, 349.

Reed, K. E., 1997. "Early Hominid Evolution and Ecological Change through theAfrican Pilo-pleistocence,"Journal of Human Evolution, 32, 289-322.

Sandyk, R., 1999. "Relevance of the Habenular Complex to Neuropsychiatry: A Review and Hypothesis," International Journal of Neuroscience, 61, 189-219.

Santamarina, E., 1958. "Melanin Pigmentation in Bovine Pineal Gland and Its Possible Correlation with Gonadal Function," Journal of Biochem. Physiol., 36, 227-335.

Skelton, R. R., and McHenry, H. H., 1992. "Evolutionary Relation ships Among Early Hominids," Journal of Human Evolution, 23, 309-349.

Tielens, A. G. G., Seab, C. G., and Hollenback, D. J. Abj., 319, L109.

Tobias, P. V., 1987. "The Bones of Homohahilia: A New Level of Organization in Cerebral Evolution," Journal of Human Evolution, 16, 741-761.

Van Kerekhoven, C., Tielens, A. G. G. M., and Waelkens, C., 2002. "Nanodiamonds around HD 9704B and Elias 1," Astronomy and Astrophysics, 384, 568-584.

Vaughan, G., Pelham, R., Pang, R., Loughlin, L., Wilson, K., Sandock, K., Vaughan, M., Koslow, S., Reiter, R., 1976. "Nocturnal Evaluation of Plasma Melatonin and Urinary S hydroxyindolencetic Acid in Young Men ," Journal of Clinical Endocrinology Metab., 42, 752-764.

Vigh, B., 1975. "Comparative Ultra Structure of Cerebrospinal Fluid Contacting Neurons and Pinealocytes," Cell Tissue Research, 158, 409-424.

Vigh, B., 1977. "Special Dendritic and Axonal Endings Formed by the Cerebrospinal Fluid Contacting Neurons of the Spinal Cord," Cell Tissue Research, 183, 541-552.

Vigh-Teichmann, I., 1980. "Comparison of the Pineal Complex and Cerebrospinal Fluid ," Fuesch. Leipzig, 94, 623-640.

Welsing, F. C., 1990. The Isis Papers: Keys to the Colors. Chicago: Third World Press.

West, J. A., 1993. Serpent in the Sky: The High Wisdom of Ancient Egypt. Wheaton, 11: Theosophical Publishing House.

Wimby-Jones, R. A., 1982. "Commentary and the Translation of the Right Panel of the Second Shine from the Tomb of Pharaoh Tutankamen," private communication. The Kemetic Institute, Chicago.

Wood, B., 2000. "Review, Human Evolution: Taxonomy and Paleobiology," Journal of Anatomy, 196, 19-60.

Zatorre, R. J., and Kruahansi, C. L., 2002. "Mental Models and Musical Minds," Science, 298, 21-39.

Thank you for the fabulous support of the Council of Elders to the Km Wr, Inc., and to Dr. John Henrik Clark and Dr. Yosef Ben Jochannan; Eye of Heru study group of Detroit; Aquarian Spiritual Center of Los Angeles; First World of New York; ASCAC, Dept. of Black Studies, San Francisco State University; Fanon R & D Center, N.l.M.H.; Black Psychiatrists of America.

CHAPTER 5

FIRE ATOMS IGNITE INNER VISION
T. Owens Moore, Ph.D.

INTRODUCTION

Given the time period before microscopes and telescopes were developed, it is amazing how ancient melanin-dominant scientists from Kemet were able to have knowledge on microcosmic and macrocosmic phenomenon. Contemplating ideas from a minute atom to planetary events is perplexing. In fact, it is only perplexing because we think today we have a monopoly on knowledge. We think we have reached the pinnacle of knowledge acquisition and anything prior to a Eurocentric way of thinking is primitive and irrelevant (Asante, 1999). The authors in this book think otherwise, and therefore, will not spend time trying to prove which age of learning was superior or inferior. The demise of Eurocentrism is evident (Asante, 1999), and the academic landscape is changing. In this book, we have attempted to emphasize that ancient knowledge was liberating, and we must return to this source of wisdom to influence a change in our consciousness.

Your two eyeballs are physically constructed to detect and process light from the external world. Without eyeballs, humans would be blind to the external world. What is interesting is that both outer vision and inner vision go to the brain for processing. Therefore, there must be overlap in the way outer vision and inner visions are perceived in the human experience. In this brief chapter, we are: 1) building on the conceptualization of Dr. Richard King's work; 2) connecting ancient knowledge to contemporary science; and 3) highlighting the power of sunlight and darkness.

Throughout history, melanin-recessive people have visited Kemet. From the intellectual theft that has occurred over time, it was the Golden Age of Greek Philosophy (640-322 B.C.) that has left an indelible mark on our current consciousness. During ancient times, many melanin-recessive invaders/interlopers/foreigners were initiated into the Mystery System, but it does not mean they truly understood the meaning of the experience. The same way we see today where cultural experiences influence behavior, we can see how African/Black music, European/White vernacular, Asian collectivity, and/or aboriginal thoughts about living in harmony with nature demonstrates that cultural experience can influence your thinking; everyone does not see the world the same. As the knowledge received from these melanin-recessive invaders was disseminated to Europe, information as to its origins was lost or used out of context. Eventually, there arose a disconnected approach to life and knowledge as a series of "specialties" replaced the "whole" view. Therefore, the misinterpretations of issues related to blackness have been contorted. We call pigment, melanin. This is a Greek word. We call the darkness of the universe that keeps the universe together, dark matter. These are Eurocentric interpretations. Those were not the concepts described by ancient thinkers, but we are forced to use those terms until African-centered intellectuals changed the discourse.

One influential scholar who helped to change the discourse was George G.M. James. He wrote **Stolen Legacy** (first published in 1954), and his book has brought many of these issues to the forefront. In his research, he provided documented evidence to demonstrate that Socrates, Plato, Aristotle, Pythagoras and Democritus were examples of foreigners who studied in the Land of the Blacks (Kemet). In Stolen Legacy, James devotes a significant portion to science principles related to the atom on pages 65 and 66.

Concisely, James explores the life and teachings of Democritus (420-316 B.C.). Democritus is the notable Greek philosopher who is associated with the science of atoms. Under HIS DOCTRINES, for example, there is vivid discussion on the nature of atoms. The section is attributed to Democritus, but we really need to understand how the interpretations of ancient knowledge are subject to the mindset of the individuals making the interpretation. Therefore, if we know Greek philosophy is stolen knowledge from melanin-dominant thinkers, we have to understand the limitations of what we read today are many times the "interpretation" of ancient knowledge. An interpretation does not equate to the truth.

FIRE ATOMS

We use the word atom in our lexicon today, but it was a conceptualization from melanin-dominant scientists. For example, the description of the atom, the qualities, the creation and the cycle of life and death associated with an atom are astonishingly similar to our understanding today. As written by Democritus and his interpretation, life and death result from a change in the arrangement of atoms. When they are arranged in a certain way, life emerges; but when the arrangement is changed to another way, then death is the result. In a related concept in modern physics, we also know that matter cannot be created or destroyed, but can be rearranged.

The Doctrine further explains that "in death, the personality disappears, the senses also disappear; but the atoms live on forever. The heavier atoms descend to the earth: but the soul atoms, which are composed of fire, ascend to the celestial regions, whence they came" (James, 1988). We are a microcosm of the macrocosm, and it is taught that we are made up of the elements of the earth. When we die and transition, our Ka or body may go back into the earth, but our spirit lasts forever and cannot be destroyed. What James has revealed to us is important as we interpret the biblical phrase, ashes to ashes and dust to dust. On the same pages of **Stolen Legacy**, the passage below provides the basis for this chapter.

The Atom is Sensation and Knowledge.

(a) The Mind or Soul is composed of fire atoms, which are the finest, the smoothest, and the most mobile. These fire atoms are distributed throughout the whole universe; and in all animate things, and especially in the body, where they are found in the largest numbers.

(b) External objects constantly give off emanations or minute images of themselves. These in turn impress upon our senses, which set in motion our Soul atoms, and thereby create Sensation and Knowledge (James, 1988).

An African-Centered Interpretation of the Doctrine of Democritus

From the vantage point of this author, passages (a) and (b) fully encapsulate how we view the topic of melanin today. As it begins with describing the Mind or Soul, we can reflect on the first Hermetic Axiom of Tehuti/Hermes – The Principle of Mentalism (Chandler, 1999). Mentalism pertains to the All is mind; The Universe is Mental. This principle explains the true nature of energy, power, and matter, and how these are subordinate to the mastery of the mind. Therefore, we can go beyond the described composition and further express what is meant by fine, smooth and mobile fire atoms that make up the mind or soul. It is proposed that melanin is found everywhere in the universe and the dark matter (i.e., melanin) described in the cosmos pertains to fire atoms.

If it describes that fire atoms are throughout the universe, in animate objects and the human body, we can speculate that the pigment melanin is the prime candidate being described. It even states that the human body is where the largest number of fire atoms can be found. Ancient melanin-dominant scientists were visualizing the experience of the original man/woman who possessed a beautiful hue on the skin. As reported by King (1990) and his reading of the ancient texts, the Flesh of Re (Ra) is skin melanin. When know Ra is the name given to the sun, so we speculate that fire atoms equate to the Flesh of Re. The connection between the sun and skin promulgates the transference of solar energy into the human body. The sun is alive in the human body as flesh. When we submit to nature and live in harmony with the elements, we naturally adapt and the ancient wisdom reflecting on this original solar power is significant. We now see the significance of the steroid hormone called soltriol which is incorrectly called Vitamin D (Moore, 2002).

According to King (1994), "the importance of skin was so important to Kamites, whose skin was Black from high levels of eumelanin, that the proper treatment of the skin was specifically listed as one of the 42 Negative Confessions for development of the Heart, Will, and Right Cortical Consciousness." In descriptions of the Flesh of Re uncovered in ancient texts (Piankoff, 1954), King makes it contemporary. He describes sunlight energy passing through melanin superconductor doorways as a flow of electrons throughout the whole body to affect the rhythmic release of neurohormones that can influence dream-trance states. Melatonin, primarily from the pineal gland, would be a candidate for the rhythmic release of a brain chemical that can dramatically influence behavior.

Melanin radiates and conducts energy. In the aforementioned passage from the Doctrine of Democritus, the external objects and the constant giving off of emanations pertains to melanin as a conduit between the material and spiritual planes of existence (Moore, 1995; 2004). These fire atoms impact our senses and this leads to the nature of why darkness matters and the power of melanin in the brain. The passage concludes that the senses are engaged and to set into motion our Soul atoms (fire atoms) or melanin. Consequently, the fire atoms help to stimulate the creation of sensation and knowledge in the human experience. The fire atoms are the source of our consciousness and our thinking is impacted by a functional melanic system. Other African-centered scholars (Clark, McGhee, Nobles and Weems, 1975) have explored the mystery of melanin decades ago. Similar to what is interpreted by Democritus, Clark et al. have expressed that both consciousness and sensitivity are at the core of black intelligence.

In sum, King never mentioned these fire atoms, but what he did recognize and reiterate throughout his writings was that the Black Dot was the seed and archetype of humanity. He expanded his work with a three-part series in his seminal book (1991), and King has been at the forefront of reclaiming knowledge pertaining to inner vision. His massive contributions should be honored as a timeless extension of ancient Black minds who initiated the first concepts thousands of years ago. May King's soul rest in peace in the ancestral realm.

Inner Vision

How do we actually obtain inner vision? First, we must be in tune with nature and the cycles of the cosmos. There is no truer source of knowledge than nature. The sun rises and the sun sets. In the darkness of sundown, we should be winding down to tune into our inner vision. In society today, we no longer sleep and rest when the sun is down because we have artificial light to keep our biological clocks out of synch with the endocrinology of sunlight and darkness. This desired balance is emphasized because ancient melanin-dominant scientists living in the pure sun during the day probably rested properly at night. Probably is stated because more than likely night lighting after sunset in ancient times was not the same level of artificial light we have today. Compared to ancient times, life and the processing of energy was so much different than today.

From reading and studying history, there was a type of energy that was harnessed thousands of years ago. Modern scientists are speculating and still trying to interpret how this universal energy was harnessed (Toth and Nielsen, 1985). It was probably not dangerous in the past, but the use of 5G for cell phone usage today is likely more detrimental. The research on the health effects of 5G towers and the potential hazards will be revealed in the coming years. Unfortunately, with the destruction of libraries and the pillaging of tombs and temples and the deliberate burning of ancient texts, we do not have easy access to many ancient secrets to reveal how energy was harnessed. How ancient people living on the continent of Africa and all over the planet were able to create a vision, construct their civilized societies, and build massive structures without the equipment we have today is bewildering. This chapter proposes that there were fire atoms that energized the highly melanized huemans to harness the necessary energy to move massive objects. When you get your mind or nous to do what you want, you can create what you want. However, it takes inner vision.

You can make yourself healthy and you can make yourself sick. You can lift up a massive object if a loved one is trapped. You can pass that test or you can fail. You can be the fastest, most agile and complex athlete on the planet or you can underestimate your talent and never rise. In other words, we must find ways to activate our fire atoms to once again harness the power of the universe. Henceforth, our inner vision will solve many of our personal issues. You can accomplish what you will, and it all starts in the brain in the darkness of neuromelanin.

Inner vision requires an exploration into darkness. We know light is critical to see and visualize during the day, but there must be physical elements in the human body, specifically the brain, that can process inner vision. Bynum has thoroughly explored these topics in his scholarly work (1999; 2012), and King (1980; 1990; 1994) has introduced us to the Black Dot and the Amenta nerve tract. In Chapter 2 of this book, Brown has emphasized the importance of melanin in neurogenesis, and each scientist has given a perspective on why darkness matters.

It is the pigment melanin that exists in the universe that harnesses the fire atoms to ignite creativity. It is a biological advance to have pigment externally in the skin as well as the internal organs. For our discussion in this book, the brain is emphasized. We will not repeat what others have already written, but we will build upon why darkly pigmented cells are the key to a highly sensitized sensory/motor network as discussed in Chapter 1.

Brain/Pyramid Connection

Inside the brain are hidden chambers and corridors in which cerebrospinal fluid (CSF) flows. Depending upon the brain site, the fluid is carrying neurochemicals and/or fire atoms that affect various regions of the brain. The CSF is flowing through cavities or openings in the brain, and we call these corridors leading to chambers the ventricular system. Similarly, deep in the pyramids of the Giza Plateau in Kemet, there are dark corridors that lead to special chambers. There are other pyramids that could be discussed, but Dr. King has written extensively about the King's chamber in the Great Pyramid of Khufu. The walls and ceiling of the King's chamber are black granite. There are multiple dark passageways in this massive structure, and the placement of a large rose granite sarcophagus in the King's chamber is intriguing. Furthermore, it is important to reveal that a BLACK pyramidion or the Ben Ben capstone was strategically placed on top of many ancient pyramids. The blackness is serving a purpose to harness energy from the cosmos.

The power of comparing the construction of pyramids and the structure of the brain can significantly enhance our understanding of how energy can be harnessed and manipulated. Furthermore, the caves located in the earth have been here on the planet for millions of years before humans existed and well before a pyramid was conceptualized by humans. Moreover, the energy in a crystal cave in the earth is how we can visualize the functioning of the human brain and the pyramid that was created by the human brain.

Personally, I have visited Kemet after the transition of our co-author, Richard King. After walking through the temples and trekking inside of the Great Pyramid of Khufu, I have a better understanding of what Dr. King emphasized in his literature and lectures. Even before King's contributions, James wrote **Stolen**

Legacy as an informative document on ancient Kemet. Actually, seeing where all of these ideas emanated on the continent of Africa provided a profound experience. For example, I sat in the middle of the King's chamber of Khufu with my modern compass on my cell phone and revealed that each of the four square walls was directionally perfect in the north, south, east and west direction. On the west wall was the empty rose granite sarcophagus that weighs as heavy as a small modern car. Why was it there? What was the purpose? How was this heavy object placed deep in the middle of this gigantic construction?

Notably, when approaching the King's chamber, there was a separate area that was symmetrically cut for you to stand in, but you had to stoop under a three-foot ceiling for about 20 feet before you got to the inner chamber. You could comfortably stand and chant in this location, and the reverberating echo was acoustically pleasant. It was a crucial aspect in the initiation rites of the elders and ancients as pointed out by Dr. Bynum in chapter 6 of this book. Chanting is used to harness energy and there is evidence that this pyramidal structure provided energy to the surrounding community (Toth and Nielsen, 1985). Modern technology gave the tourists electricity and modern light bulbs to walk the corridors, but what energy source was used during ancient times to see inside these structures when the pyramids were built?

Those builders who constructed this massive power generator probably did not intend for us to be inside. Many of the passageways into the pyramid were originally covered, and even today, heat absorbing radar technology has revealed additional corridors and chambers that have yet to be opened for investigation. As a neuroscientist, there are also uncharted territories in the human brain. For example, the pineal gland is connected to the brain by the habenula. From the habenula, there is a pathway leading down the brain to the interpenducular nucleus at the base of the brain near the hypothalamus. The pathway is the stria medullaris retroflexus. Modern science has not revealed the significance of this pathway or corridor.

In addition to reflecting on passageways in the human brain, chants using the acoustical term Om/Aum can stimulate the ventricular system with vibrations. These vibrations can cause the release of chemicals in the brain's "King's chamber" where the pineal gland resides. The darkness of this third ventricle bathes the pineal gland deep in the middle of the brain. Similarly, the King's chamber in the Great pyramid is in the center, and this King's chamber is the macrocosm of the microcosm in the brain where the pineal gland is located.

CONCLUSION

In sum, in the corridors and passageways deep in the recesses of the microcosmic brain and the macrocosmic monolithic structures are where visions are constructed. It is into the dark where light forms into a penetrating form of consciousness to help formulate an idea. We all have ideas, but how do we make these ideas manifest in our life. This chapter suggests that the energy of fire atoms stimulates the creative mind when we meditate. Fire atoms are working during improvisation and creative experiences. Proper meditation helps to activate those internal structures in the brain. When we combine meditation inside a grand energy source such as a pyramid or other ancient temples or a cave in the earth, there is a tremendous energy to ignite visions. Therefore, tune into your fire atoms so you may tune out of nonsensical thinking so you can tap into the divine universal force.

REFERENCES

Asante, M.K. (1999). The Painful Demise of Eurocentrism. Trenton, NJ: Africa World Press, Inc.

Bynum, E.B. (1999). The African Unconscious. New York, NY: Cosimo Books.

Bynum, E.B. (2012). Dark Light Consciousness: Melanin, Serpent Power, and the Luminous Matrix of Reality. Rochester, VT: Inner Traditions.

Chandler, W.B. (1999). Ancient Future. Baltimore, MD: Black Classic Press.

Clark (X) C., McGhee, P., Nobles, W. and Weems, L. (1975). Voodoo or I.Q.: An introduction to African psychology. Journal of Black Psychology, 1 (2), 9-19.

James, G.G.M. (1988). Stolen Legacy. San Francisco, CA. Originally published 1954.

King, R. (1980). Black Dot - The Black Seed: The Archetype of Humanity. Uraeus: The Journal of Unconscious Life, 2(1), 18-23.

King, R. (1990). African Origin of Biological Psychiatry. Germantown, TN: Seymour-Smith, Inc.

King, R. (1994). Melanin: A Key to Freedom. Hampton, VA: U.B.& U.S. Communications Systems.

Moore, T.O. (1995). The Science of Melanin: Dispelling the Myths. Silver Spring, MD: Beckham House Publishers.

Moore, T.O. (2002). Dark Matters Dark Secrets. Redan, GA: Zamani Press.

Moore, T.O. (2004). The Science of Melanin (The Second Edition). Redan, GA: Zamani Press.

Piankoff, A. (1954). <u>The Tomb of Ramses IV</u>. Bollingen Series XL-1. New York, NY: Pantheon Books.

Toth, M. and Nielsen, G. (1985). <u>Pyramid Power</u>. Rochester, VT: Destiny Books.

CHAPTER 6

The Farther Reaches of Eldership and the Dreamlife of Families.
Edward Bruce Bynum, Ph.D., ABPP

THE ANCIENT LINEAGE AND MODERN TIMES

Within the dreamlife and culture of African American families, as well as traditional cultures all over the earth, dreaming on a deeper level has always been associated with knowledge and wisdom. This chapter will explore this as it pertains to neuromelanin, the material in neural tissue that we believe conducts a form of bio-light itself and many subtle dynamics of the brain and human culture. Throughout this book we have been suggesting in different chapters that melanin and neuromelanin in particular is implicated in the central control of numerous biological, physiological and psychological processes. It is really *a multireceptor* of a full range of electromagnetic, acoustic, visual and other more subtle fields. Because of its deep affinity with light and bioluminosity we believe it plays an integrative role in transducing matter from form to form within our physical bodies and similarity with our deep psychological functions. One of those psychological functions is wisdom and the kind of information sharing often associated with those who are venerated in traditional societies and cultures, our elders.

The wisdom of our ancestors was often believed to come through our dreams or their dream interpretation. While bits of data and new ideas were and are often seen as the province of youth, distilling that knowledge into wisdom has for untold millennia been the domain of the elders in these societies. The ancient Romans, Greeks, Kemetic Egyptians and West Africans all venerated those who came through youth and adulthood into an older age intact. It implied a hard earned insight into the workings of the world. The Romans codified this and these 'seniors' formed the basis of the Roman "senate" in both the early republic and later imperial age. Greek images of their elders such as Socrates, Plato and Aristotle even today are the idealized images of wisdom in the Western world. Similarly with the Chinese image of Confucius.

West Africans and others on the continent have various images of the wise elder in the diviner and the Orisha and even of one's own individual spirit represented in one's head or Ori. Traditionally this is what it meant to be "strong in the head". In these traditions when the spiritual- energy or "ase" is made to rise up through various means and techniques to meet with the "ori" in the head, we have 'ori-sha' or the *Orisha*. This is very similar to the dynamic of the original south Indian or Dravidian yogi's energetic awakening and rising of spiritual energy or 'Shakti' up along and through the spinal line into the head or Shiva in the process of enlightenment (Bynum, 2005). In the West African disciplines and traditions elders usually accompanied the neophyte through the mysterious processes of initiation. This was not only because of the need to know the secrets of the rites, but also because the deep unconscious was encountered with all its awe and terror which could be a rough ride and demanded explicit training. Elders were believed to be "strong in the head".

The ancient Egyptian initiate into the temple mysteries was often conducted through their training by these venerated elders and priests, adepts who were skilled in both the language of dreams and the physical sciences and medicine of the time. They were held to be the "stable ones", the wisdom keepers. Given the extensive art of mummification practiced and documented in their texts over thousands of years, plus the widespread use of geometry and trigonometry required to construct and raise the obelisks and pyramids, they clearly knew a great deal about anatomy, chemistry, engineering, mathematics, astronomy and also the workings of the mind on many levels.

They had two names for what we conceive to be the unconscious region of the mind today. They termed it the all black underworld of the *Amenti* or *Amenta* and also the *Primeval Waters of Nun*. Here the world itself arose out of the depths of reality into manifestation in space and time (King, 1990; Hourning, 1986)). Their concept of the unconscious was much larger than ours today and transcended not merely the personal but also the collective unconscious and from there went on to touch upon the *superconscious* level of mind. As the original Tehuti and his later Greek interpretation as the 'Thoth' figure, the primogenitor of the Hermetic tradition observed, "All is Mind" (Chandler, 1999).

UNDERSTANDING DREAMS IN THE LARGER UNCONSCIOUS

In the West African disciplines the closest approximation to our present day clinical understanding of the unconscious is the *Ayanmo* concept (Morakinyo, 1983). However, that notion is more intimately associated with the existential choices we make in life even though on many levels these choices are not fully known or conscious to us. Notions of fate and destiny also enter into the equation along with spirituality and our family dynamics. All of these are believed to have their dynamic expression in the dreamlife with its consequent influence on the mind and body. It was often an elder during initiation that assisted the new person, the neophyte, in this expansive process of understanding.

Some dreams are older than we are and can appear, almost unchanged, from generation to generation in our families. This is seen in some family events when a shared idea is passed on and on. When it comes to society at large the power of an uplifting dream can be seen to unite large families and even larger groups of people in society. Those immigrant families from the Caribbean that literally dream of money, work and 'freedom in America' are part of our national fabric and consciousness. "Next year Jerusalem" has been recited over religious holidays and united Jews for centuries. Martin Luther King's "I have a dream today" speech still echoes as one of the most magnificent in American history, not only for the political aspirations it drew from but also the idealism and visionary impulse it unfurled across the generations in America regardless of ethnicity. Each was a multigenerational transmission of an essential dream rooted in the personal that moved out into the collective experience of people.

When we look at families as we dream about specific events such as marriages, pregnancy, deaths, substance abuse problems, medical illnesses and so on, it is interesting to notice how the very young often come to symbolize vulnerability and the future and how we ourselves are *actively incorporated* as images into the intimate and interior dreams of our other family members and friends as symbols. Our actual image in our families changes and matures as we evolve from infant to elder over the epigenetic stages or lifespan in our family for better or worse. We are *enfolded* into and share *in-form-mation* with these special people in our lives with us having a somewhat inverse square representation of ourselves in the other. We literally live in the inner landscape of our intimate others, for better or worse. This is the familial unconscious (Bynum, 2005). Our image and roles in the lives of others are often complex, emotionally charged and perplexing all at the same time. Our dreams reflect this in the *shared dreamlife* of families (Bynum, 2017). It is reported that some 20% of our dreamlife in one way or image or symbol or another involves a member of our own family. In the familial unconscious there are "what's" going on but also innumerable "whos" that are doing "whats".

Actually, the whole field of family therapy in modern psychology is *tacitly* based on this very notion that family consciousness is as deep, if not deeper, than our individual consciousness and identity, literally a family unconscious. In family therapy the clinician implicitly asks family members, in the African and Asantian sense, not "***what*** is the matter with you" but rather "***who*** is the matter with you". The universe is pervaded by personhood or *Personalism.* Living forms, persons, even inanimate objects are imbued with this *essence* which is held to be as fundamental as matter and spirit.

Now biologically speaking we know that REM sleep or coordinated Rapid Eye Movement in sleep is generally associated with dreaming. Dreaming is sometimes complicated by depression or illnesses like Parkinson's. Some medications we take compound these situations. It is also known that REM is associated with certain kinds of intense dreaming such as lucid dreaming where subjectively we become self-aware when we are dreaming and, with technique and discipline, can consciously influence actions in our dreams. However, it is also associated with the West African shamanic discipline of bodily rigidity and subjectively experiencing the sense and sensation of "flying" in our sleep. This is when it is believed "the witches are riding me". It has a powerful arousal effect on and within the entire body, psychologically and somatically. This contribution to the science of consciousness has yet to be explored in modern psychology, in either traditional European or even Afrocentric circles. We sense that in these and related rites and practices untold millennia old in our African generated species are the Black origins of mysticism and psychology (Bynum, 2021). Because of its origins even to this day it remains shrouded in taboo. Well trained elders of the past in touch with their root traditions however would explore this with the student or initiates to great advantage in their psychological and spiritual growth.

The Dynamic Legacy of Eldership

The Great pyramid of the wind-swept Giza plateau in Egypt has built into it shafts that seem to be designed to capture and funnel the light and *resonance* from specific stars and constellations at designated times in the solar and Great Year. It was not, as supposed by early European scholars, the final resting place of the pharos, i.e., in so called king's or queen's chambers (Hancock and Bauval, 1996). No. These chambers, the sarcophagi and adjoining rooms, were places of initiation into the solar mysteries and the symbolic rites of resurrection facilitated by the skilled use of hypnosis, autogenics, lucid dreaming and rhythmic verbal incantation of certain select tones and sounds *that would induce clinical dissociation* and OBE or out of body experiences (Edmonston, 1986). Remember these ancients, these "stable ones" and keepers of wisdom, were well aware of the dynamics of the unconscious and functionally gave it a name, i.e. 'The Primeval

Waters of Nun' and the 'Amenti' or 'Amenta' (Hourning, 1986, King, 1990). The royals were initiated into the mysteries here yes, but they were buried with their families in rock cut-out underground tombs and mastabas in the Valley of the Kings. Not in the great pyramid or the other two of the Giza plateau! Of the roughly 138 or so pyramids in the Egyptian region very few full or authentic mummies have been found. Of the few remains that have been found such as at Saqqara and Menkaure's at Giza they seem to have been interred during *later* periods. 'None has been found in any of the huge Golden Age pyramids' (Creighton and Osborn, 2012).

Make no mistake about it, the purpose of these rites using, these various methods, was to induce different forms of what we call today 'clinical dissociation', including hypnosis, and also 'OBE or out-of-body experiences, i.e., *autoscopic perception*. This included the ritualistic 'lying on of hands' and utterance of sacred words kept hidden from the uninitiated (Bonwick, 1878). These included and induced for sure the vivid sensation of lucid or conscious dream travel. The initiate, following specific preparations and discipline, would lie in the sacred sleeping room. The disciplines were applied. The initiate experienced leaving the physically embodied condition, rising like a bird (*Ba*) or butterfly, and flying about in or ascending briefly into what we might describe as intelligent higher order forms of enfolded light, thereby fulfilling the perception and intuition of forms of life not usually witnessed in the daily waking states of consciousness. They experienced being visited by a 'god' such as Isis or Serapis and specific, highly personal information(in-form-mation) was passed on to them in some form of resonance. The initiate left these rites feeling embodied by the gods, a literal 'garment of Isis' or other deity (Ozaniec,2022). This was in preparation for the eventual fusing of one's consciousness with the perceived 'bird of light' that arose to the initiate at the beginning and during their post-mortem passage.

These elders believed that wisdom encoded within and beneath the pyramids would rise up to the initiate in trance and that a higher insight would descend into the soul during these rites of passage.

We might remember here that often 'symbolic' or mythological experiences are expressions of actual psychoenergetic exchanges and processes. The psychodynamic identified Oedipus complex observed in certain familial configurations is a real dynamically experienced pattern of energy exchange. Our labeling it Oedipus or other 'mythological' name is merely our surface name for an enduring pattern of human experience.

During these rites the initiate had a direct experience of what the ancient Egyptians referred to as the Kaba, or 'ghost' that surrounds the dense physical self of matter. It was in a sense a subtle body, a kind of quantum field body or envelope of light if you will, capable of memory and experience of a different order than the denser physical body in which it was partially housed. Remember that just as there is light above and beyond the body in the wider cosmos, there is also a sea of light energy *below the quantum* surrounding and infusing the body. Human consciousness in certain vibratory states appears to be able to interact with both.

The initiate, by participating in this, went far beyond theory and *experientially validated*, at least briefly for themselves, the separation of mind from body, experiencing the actual *mutability* of time, matter, space and gravity. There was a deeper resonance that was experienced to underlie ordinary matter, space and time. Yes time and space are mutable at very high speeds as in Relativity theory, but also in dreams and certain transcendent mystical states.

We certainly *live* in space, but we also *express* space; we also live in time, but we also *express* time. All of these events and experiences demonstrated unequivocally to the initiate the deeper resonance of the soul's existence. At some point, with sufficient discipline and practice, the nonlocality of consciousness itself was personally realized and the experience of what we might term trans-dimensional travel appeared possible (Bynum, 2012).

While this occurred within the chambers and upon rising from the sarcophagus, it *simulated* death and rebirth, but was not technically speaking a classical NDE or near death experience. Yes it was to enter the anteroom of death and briefly the 'bardos' that followed, and yes there was encountered a great being of light or 'light-bird', judgement/Maat, etc, and a kind of 'life-review', including how their lives were interconnected with other lives on earth in an intimate loom. But the bodily form itself was retained and the transformation of consciousness that was initiated was returned to a human form. There was no ultimate *translation* into something else, no actual lasting fusion or complete identification with the light bird, unless one had completely mastered the deeper disciplines and teachings. We will return to this notion in the latter sections of this chapter.

In these moments of inner illumination the initiate experienced a clear perception of human souls in great cycles and sub-cycles of birth, death, and rebirth, migrating to earth from subtle regions beyond their localized three dimensional plane. In this process of evolution there were perceived to be progressive, albeit non-linear, movements bringing progressively more good into the earth as the soul evolved. This was the intuition behind the Kemetic Egyptian practice of mummification, i.e., that the soul would travel and learn after death, then return and reincarnate into denser matter again and be elevated higher and higher though spiritual guidance and discipline.

Others believed the soul left the body at death but then returned to the bodily form endlessly to new itself, thus the need to keep the mummy intact. It was also believed that a physical image or sculpture of the deceased could serve the same purpose as long as the nose was left intact to receive in and then release again the vital breath. By defacing the nose this process was believed to be stopped, thus the ritual of breaking the nose of the statue.

This notion of cycles of embodiment and reincarnation pulsing from a vast transcendental Intelligence is also at the root of many West African religions, perhaps most notably in the IFA tradition with its overarching Orunmilist vision and philosophy. Indeed not only do human souls 'reincarnate', but so do whole nations and waves of peoples and perhaps eras (Akiwowo, 1980). By learning to dissociate briefly from the body in various ways from the dense localized four-dimensional world by specific disciple and technique, then entering the fifth dimensional realm of light, this perception was made possible.

Dream fragments arising out of their experience at this time would have appeared to come from not only their present life and time, but perhaps fragments from other lifetimes in other tribes and on other regions on earth and perhaps beyond. The 'arrow of time' was not limited from past to future but conceivably from future to present, as nonsensical as that might seem to us today. The initiate was able to 'see' themselves in the past and also 'remember' themselves from the future.

In the quantum domain the 'arrow of time' and of causality is not always clear. This notion of reincarnation may seem remote, but there are actually hundreds of extensive and closely investigated field studies of this phenomenon by well-trained observers in the scientific method that our present era tends to 'selectively in-attend to' because they do not fit neatly into the prevailing paradigm of matter, energy, psychic and identity boundary formation. Nevertheless the data is vast and clinically strong, supported by medical records, physical markings on the body and other biological information. (Shroder, 1999; Stevenson, 1997; 1995; Mills, 1994).

Elders and priests had been experimenting with these methods for untold millennia before the rise of the pyramids. On the great grasslands, in dark caves and grottoes, and in the recesses of the dense forests, elders had no doubt noticed spontaneous episodes of dissociation and experiences of the sensation of 'travel' but it took millions of hours of experimentation before the early formulas and rituals were worked out into methodologies.

Civilizations and cultures came and went. The experiences multiplied and deepened. Eventually this intuition of an inner light and the actual externally perceived light of the stars coalesced into a sense of an intimate connection between the two. From these early roots a transcendental psychology was born.

The perception of an internally luminous dimension correlated with an externally perceived pattern of light above lead inevitably to the early intuitions of star and soul patterns or *correspondences* that formed the root of the knowledge systems of the early astrophysicists along with their parallel beliefs in astrology. This was the case among peoples as diverse as the Dogon of Mali and those who practiced the Mithraism Rites of the Roman Empire as well as others who created the archaeo-astronomical structures scattered in ruins left by early peoples around the earth. These were often quite accurate star and celestial pattern mappings and throughout the megalithic age were evident from isles of Britain to Nabta Playa near present day Sudan. Many scientists are perturbed at the accuracy of these maps and often dismiss them because there is no known basis for their representation in terms of traditional scientific epistemology. How on earth they ask could these rites and their associated divination rituals accurately reveal patterns in the universe. However we are suggesting that neuromelanin's subtle bio-luminous qualities intimately associated with the human nervous system is the leading scientific candidate for the biological basis for this perception or *introception,* of internally luminous phenomena. It may well have made this possible in a methodology we have lost. This has not only biological and neurological but electromagnetic and even quantum mechanical implications. Actually this implication of *a quantum potential* opens the door to aspects of wider connectivity and nonlocality more traditionally explored in experiences of deep meditation and states of religious expansion. These rites, calculations and divination rituals were often rooted in the base 2 calculation system, not 10. The base 2 system later gave rise the calculation systems we use in much of science today. More on this later in the section on eldership, divination and your local computer laptop.

So as phenomenal as it may seem to us today the underlying science of consciousness embedded here allowed for communion with what were believed to be the nonlocal intelligences and the stars. The initiate realized, with the help and direction of these elders and priests, that they were on a journey that passed through many lifetimes as the soul ascended into a realm of light after its mythic fall for unclear reasons ages before into the density of matter and time. It was to realize the high ancestral realm and far beyond. The epigenetic cycles and sub-cycles were clearly stages of the soul's adventures on earth and were on full display to the initiate. Their life was forever changed. The knowledge imparted each cycle was meant to educate the initiate as to their deeper real identity beyond matter and time and to pass this knowledge on to future generations who were deemed worthy and ready.

As we explore these reported experiences, we are gently reminded that most documented experiences of the kinds mentioned above are in contemporary times viewed through the constricting lens of mild or gross psychopathology. Our human capacity for various forms of what we label 'dissociation' has unfortunately been largely framed by modern psychiatry and clinical psychology. What was classically perceived as the 'soul' of an individual capable of travel, inner vision, and even recall of deep and primordial memories has been cast today as misperception, hallucination, or worse. Modern clinical theory has thrown the baby out with the bathwater. Because many who do report these experiences, actually are clinically compromised in some fashion does not mean all who report or experience these are in the same compromised situation.

It is the human capacity for nonlocal consciousness that undergirds these phenomena, whether they are healthy and functional or not. Both involve forms of separation of mind from body, anomalous perception and shifts or losses in memory. In regular clinical dissociation assessed in the clinical office or hospital one can have lost memories to various degrees, but they are all of *this life*. In some of the other experiences, we have been describing however it is not so much the part of this life that have been lost, but the memories of another *wave of life and experience* around the core soul that had been lost and must be retrieved. But by collapsing this human capacity into a medical diagnosis we severely limit our own perceptions and knowledge of who and what we truly are. All of this is about the pursuit of balance.

Undoubtedly in these episodes during the ancient rites the initiate experienced the deep transformative processes that others for millennia before them had experienced and no doubt drew strength and courage from it. For the soul of the king or pharos in particular this was codified in the earliest religious writings we have from that period known to us today as the pyramid texts (Faulkner, 2007). They entered the same seeming river of eternal consciousness and witnessed for themselves that the soul they re-experienced was not a prisoner of space, time and matter, but, like light itself, *projected* itself through space, time and matter and dwelled in a dimension beyond these (Bynum, 2012). They understood by direct intelligence and personal revelation that indeed they were inherently beings of light in a nonlocal universe, a *multiverse* by our words today. Some promulgated doctrines of lives in other worlds, other dimensions, yet remaining intimately connected to this world. It was a quest for balance in the intimate expanse of the cosmos.

At its highest points these rites helped awaken the subtle sleeping current inside and along the spinal column referred to by many as the 'serpent power' or kundalini. The pyramid texts (Faulkner,2007) the most ancient religious writings we have that date back to pre-dynastic times were codified at least 5000+ BCE. They refer in 'utterances' to the progress of the soul toward awakening in this life and the afterlife. The elders and masters who helped raise and stabilize the pillar or column within the body, referred to as a djed, facilitated the rise of the column and its *conductivity* of the life energy, toward the 'eye' behind and above the eye, or 'third eye', which as described here in chapter 4 and elsewhere. It appears to be the photoreceptive pineal gland (King, 2001). The ancient festival of the raising of the Djed in those pre-dynastic times, times which by the way the *Egyptian Book of The Dead* was also composed, was about not only raising physical buildings, but also psychospiritually the discipline of raising the *energy* in the djed column of the spine. It was thought to occur during times of great change, the end or beginning of the new ages of humankind. *To raise the djed column and its internal energy current was to elevate the backbone of Ausar or Osiris from the lunar underworld of regeneration and rebirth, to actively assist in the ascension of the human soul.*

This Djed column of the spine was conceived of as a kind of hollow column or tunnel of light connecting the pole star of the sky during that age with the living current moving through the human spine. The pole star is that star in the night sky that, during the precession cycle, is at the still center of the sky and all the other stars orbit around it. One's spiritual work was moving up and through this column or axis. Beyond the pole star was thought to be the realm of the gods and higher vibrational dimensional beings. Today we might refer to this as the supramental or superconscious level of the mind (, 2012) The ultimate goal of spiritual disciple was to reach this point. The later Roman followers of this religion, like the Kemetic Egyptians themselves, had these djed pillars painted on their coffins to symbolize becoming as Osiris.

The Body and Connectivity to the Physical Earth

It is useful to recall that Schwaller de Lubicz in the last century demonstrated, after decades of measurement and on site study, that the exacting mathematical and geometric sequences of the temple complexes of Luxor in Egypt were laid out according to a rhythmic harmonic plan (1998; 1961). The ringing of the Egyptian obelisks may well have set the tone since select tones and sounds were believed to induce a specific flow or current or flow of energy in the gross and subtle body of the initiate and elder-priests. Later in history of course these particular tones and sounds degenerated into a kind of series of formulas that people used to less and less avail as the original high traditions died out. It is more likely that amongst the priesthood within the walls of the temples and pyramids that the more exacting tones and sounds remained intact for a longer period of time before they too died out there.

While this chapter focuses primarily on Kemet or ancient Egypt we would be remiss if we did not mention that Egypt and its monuments was one of several sites on earth, many connected with each other, where megalithic cultures precisely mapped trails of the stars, the equinoxes and other astronomical events and reflected this in huge stone monuments on the earth's surface somewhat like a celestial clock. They were highly sophisticated in the astronomical pathways of the known constellations. Stonehenge in pre-Roman Brittan and Gobekli Tepe in Turkey were also ancient sites where our ancestors practiced their sacred rites of initiation into the solar mysteries. These may well have been more than astronomical calendars laid out on the earth in precise alignment to inform us when to plant and harvest crops. We suspect they were much much more.

Perhaps alignments with the earth's electromagnetic and gravitational fields were harnessed by these structures. After all, given the four so-called fundamental forces in nature, we do actually subjectively *feel* the forces of gravity and electromagnetism in our daily experience, unlike the strong and weak forces of the atom. When these extensive megalithic structures, now largely in ruins across the earth, were active and resonate, they may have been able to enhance a *piezoelectric effect* in the human brain that was stimulated by activating rituals, rhythmic breathing, incantation and other disciplines alluded to earlier that were used by the priests and elders. Remember that the Egyptian sarcophagi used in their pyramid rites were often granite stone of high crystalline content. This piezoelectric effect occurs when a gel-like material, such as the thick gel of the brain, is made to vibrate in a contained space, converting its mechanical movements into electromagnetic waves and energy. At some point these 'waves' of energy interact with the more subtle 'matter waves' which characterize all the material world as revealed by the Nobel Laurette physicist De Broglie in the early years of quantum mechanics. These 'waves' in turn become as aspect of the wider interference patterns of the cosmos, a cosmos that is resonate, wavelike and holonomic in its dynamics.

Our ancestors, ages before the skies were clouded with human debris of various kinds and the earth was unencumbered and unpolluted, may very well have been able to sense geomagnetic and gravitational anomalies on the earth. Ceremonial sites like these are scattered across our world from the earliest of days. By shamanic ritual and other disciplines that shifted consciousness they may have been able to create or recognize portals to the nonlocal intelligences and the stars (Bynum, 2021; Collins and Little, 2022).

Neuromelanin is at the dynamic biological interface of this in the human brain and helps *transduce* energy and information from one state to another. Through rhythmic breathing, heart rate changes, chanting of select tones and sounds, incantation and other rituals, this piezoelectric effect can be achieved when the heart-aorta system is activated through disciplines and the brain is made to vibrate within the skull casing. These stone structures alluded to here are often high in crystal content, especially granite and basalt. They have the capacity to subtly *resonate* under certain conditions and perhaps in states of alignment with the brain's consciousness convey information (in-form-mation) of some form to the practitioner. We might remember that, in science anyway, it seems it is *'information'* that underlies, unites and is more fundamental in nature than even the quantum and relativistic worlds Currivan (2017).

While it is highly probable that the brain itself creates this *piezoelectric effect*, the vertebrae of the spinal line, given its cushions between vertebrae and its capacity to vibrate, may well be integrated into this process. Thereby the whole skull-brain casement and the wavelength of the spinal line could resonate with and within the larger environment.

In this way it appears that these stone and crystal structures may enter into a *coordinated resonance* with the human brain and energetically affect the nonlocal dimension of our consciousness in radical ways we still do not fully comprehend. The use of the whole process was to expedite the 'ascension', at least briefly, of the local consciousness into a more nonlocal and vaster stellar consciousness of information and insight. No doubt during the pyramid rites of these "stable ones", as well as other rites, the subtle practices of dissociation were used to facilitate the experience of 'travel' in an interconnected and perhaps holographically informed universe (Currivan, 2017) by affecting a warp in the higher geodesics of space in ways we presently can only imagine but will one day yet rediscover.

From this perspective of the soul passing through many different 'waves of lifetimes' as it unfolded, dreams and dreamlife served not only as a source of individual revelation, but also a familial and collective one in which these dreams at times acted as a kind of informational life wave transmitting data over many eras of the soul. Before their successful initiation and transition the soul was like an infant falling asleep at night and then re-awakening with only the vaguest sense of what may have happened to it the previous day. When fully 'awake' however the initiate, like an awakened lucid dreamer, was conscious and aware even while dreaming. Different lifetimes appeared like different dreams of the same dreamer before lucid awareness penetrated through to the deeper underlying identity of the dreamer. The light of the self suddenly realized it was *projected* through different tissues of space, time, matter, forms and even lifetimes.

During these literal "rites of passage" in the pyramids the techniques of unfolding or reaching the higher self in a dream-like trance involved use of some of the above mentioned psychological methods plus two in particular of the five so called 'Platonic' solid shapes. These five shapes are the physical shapes that underlie our physical dimension. In other words, all shapes of our three-dimensional material world of height, depth and width, are combinations and variations of these five fundamental shapes in nature, shapes discovered in the ancient Egyptian study of geometry.

Obviously, the pyramid or tetrahedron shape was used, but also the cube, the symbol of manifestation in time and space. Remember again, that we are not only contained *in* space, but that we also *express* space, are contained *in* time but also *express* time in our most intimate experiences. There are numerous statues showing the emergence of the human head out of a cube in ancient Egyptian religious and art work (West, 1993). The intention it seems was to demonstrate the emergence of the initiate out of the flat two dimensional plane to our three dimensional cube, then on to the four dimensional cube projected in hyperspace (see the Salvador Dali painting of the Christ on a cross) and finally to *the fifth dimensional* configuration that reflects light and mass/energy itself, an idea which was introduced into modern physics by Schwarzchild to the awe and admiration of Einstein (Bynum, 2012). This is all an ancient story.

The African American Mythogenesis

Now for African Americans in particular it is interesting to notice how language associations came to be woven and implicated in this process. Again, however, this is by no means limited to African Americans and indeed as a biological and physical phenomenon is implicated in the roots of all human beings. You see the name Schwarzchild in German translates literally as 'Black Child' and one of the most popular Black musical groups of the sixties and seventies was actually called *'The Fifth Dimension'*. Their most famous recording was a kind of initiatory anthem called "*This is the Dawning of the Age of Aquarius*".

There are many other examples. One could easily fill volumes on music alone on how this *interiorization* of a kind of African American mythos has been introduced into the wider American cultural landscape through Rock and Roll, 'soul music', the blues, each of which, like jazz, bebop, gospel and even ragtime before them all went through a process of rejection, conflicted tolerance, acceptance and finally *interiorzation* in the American and world psyche. People in Asia and Europe today create "rap music" in their own languages!

This African American mythogenesis has ancient roots and a direct line can be drawn from pre-dynastic Egypt over at least 5000+ years ago. Some, including the Egyptians own written history of themselves, traces their origins back around 12500 BCE, to Zep Tepi, the 'splendid time of the first time' mentioned and carved into the Dream Stele which stands between the paws of the sphinx (Creighton and Osborn, 2012). While it was placed there over a thousand after the events, it was a call to themselves as to their own origins, a true myth.

On the plains to the west of the Egyptian Sahara stand ruins of a remote antiquity. Archaeologists have recently uncovered human megalithic structures there that reflect an accurate astrophysical map of the stars and constellations at Napta Playa. Some markers note stars that were seen with the naked eye and some only seen by other methods as yet unknown. This includes detailed astronomical and cosmological information, such as the distance to certain stars, the speeds at which stars are moving away from us on earth, the structure of the Milky Way and other information (Brophy, 2002).

It is almost as though these ancient star watchers felt a living connection to the stars, as though the stars had implicit roots or correspondences in their own souls and brains and that they could absorb this light from so far away and bring it into the interior of their lives. With a light absorbing neuromelanin rich brain and brainstem however this notion of a *resonance* between remote external light and internal spectrums or foci of absorption may not be as inconceivable as we might think (Bynum, 2012). As a species evolving toward a sublime state of consciousness we are really still in our infancy.

Actually, we really should not be so surprised by all this since this region was home to other, later, mappers of the earth and stars. The great philosopher Eratosthenes of Cyrene (270-195 BCE), had been a native of Cyrene in southern Egypt near Aswan where the great dam of the Nile is today. He had noticed in Cyrene that on the summer solstice the sun cast no shadow at noon. By complex deduction and measurements of differences of the same sunlight-shadow relationship in the city of Alexandria, he was able to calculate by planes and degrees and the differences between the two sites, the circumference of the earth. He was only slightly off but by implication could also intuit that the earth was round. He was building on a long lineage of knowledge emerging in this area of the world and had been the librarian of the museum and library of Alexandria, itself thousands of years old. This was the same library of thousands of books from antiquity later burned to the ground by the Romans under Julius Cesar. Whether he actually 'discovered' this knowledge or inherited it from earlier sources is unknown.

How this knowledge might have arisen in the first place has been attributed to either some vast lost ancient civilization, extraterrestrial contact, or maybe, just maybe, through a pathway of human apprehension unknown to modern medicine and science but highly possible with an understanding of human consciousness interacting with the physical world through a different but still very human epistemology. In Chapter 4 of this book this was explored in the context of a kind of neurocomological holography rooted in the neuromelanin structures throughout the brain and brainstem that, under certain conditions, interact with and absorb subtle forms of light.

The civilization that discovered the earliest roots of this information either fell or morphed into the Kemetic Egyptain civilization (Bauval & Brophy, 2011). That civilization itself seems to have given rise to the vastness of ancient Egypt through the millennia in terms of astronomy, mathematics, geometry, huge architectural structures and medicine that still sings to us today. Eventually after thousands of years that civilization was overwhelmed by internal and external forces and went into decline. Migrations from there dispersed through the Nile valley to other parts of the world along with religious and scientific ideas, (Bernal, 1987; Diop, 1991,1974). The three principal religions of the western world, Christianity, Judaism and Islam each have structural roots in this history and region of the world (ben-Jochannan, 1970). For us in particular these migrations further to the west are of interest, especially through the Dogon people of West Africa who trace their own migratory history from those times to the medieval period of West African history (Finch, 1998).

They undoubtedly also had contact with other civilizations over time from the interior (Jackson, 1970). They too were known as a star gazing star mapping people with a particular affinity for the star Sirius, a star the Egyptians had known well and incorporated into their calendars and religious rites.

The Dogon knew of the roughly fifty-year spiral-like orbit of its unseen and denser binary, the 'star behind the star', Sirius –B or Digitaria along with its associated stars. This star, Sirius-B or Digitaria, was unknown to Western telescopic observation until 1862, let alone was it known as a star of denser matter. These rites have been an integral part of their religious rituals for at least 700 years. The ancient Kemetic Egyptians had known of this unseen star, expressed it symbolically in their rites and statuary in the form of the 'sun behind the sun', the feathered hawk behind the head of the Pharaoh. Again all this had not been observed by any telescopes. Many of their most secret and sacred 'dark rites' were associated with its veneration (Griaule & Dieterlen, 1986; Krishna,1978).

The Dogon of ancient Mali lived and still live in the upper Niger River area. They had been incorporated into the empire of Mali along with other peoples. The peoples of Mali were also a large water voyaging and sea fairing people and seemed to have had contact in the Americans before the voyages of Columbus (Van Sertima, 1976). Many captured and indentured Africans came from this region to the west coast during the slave trade. Many of their descendants were then trafficked to North America and Caribbean.

The free Black almanac publisher, clock maker and surveyor Benjamin Banneker, was an intimate part of the construction of the new city of Washington DC. Banneker learned a good deal of formal astronomy and geometry from the nephew of Andrew Ellicott, a leading land surveyor of the time. They helped find and clear the chosen land set aside by the Constitution of the new nation. It was these two, along with the Frenchmen L'Enfant, who by the way was forced to leave the job proposal early on by Thomas Jefferson over a financial dispute, helped lay out the streets and buildings of the early Republic of the United States. They had the *explicit* ideal of guiding the stars and their celestial energy down into the ground work of streets and buildings of the new nation (Browder, 2004). It was actually Jefferson who, on the basis of his published reputation, had suggested Banneker to Ellicott for the job.

Banneker had his ancestral roots in this region of Mali. In all likelihood he heard some of the hidden and sacred lore of the stars from his grandfather Banneker who spoke of a certain ancient wisdom about the stars to members of the family (Bynum, 2012; Cerami, 2002). This intuitive feel, not merely speculative and abstract 'artistic' conception, for the *vibratory presence of the stars* can be traced all the way down to the music of today in the spiritual jazz of Pharaoh Sanders and Afrofuturism composer Sun RA., whose influence continues on in artists like Solange as well as many others. This current of creativity is much deeper than the space we have here to explicate it but the great 'winding waterway' is clear and has been deeply incorporated into America's cultural history and consciousness.

It is also interesting to see how modern cinema touches on this. The popular film **Black Panther** has an advanced African society secretly fueled by a mysterious energy called *vibrainium* (a la 'vibrating brain') which must be kept secret or it will become dangerous in the hands of others, etc. Thereafter follows an entertaining narrative with great imagery and both cultural and subtle religious overtones about the ancestral realm. Good for entertainment for sure but the connection to the actual Black Panther political party of the 1960' and 70's in the United States is hard to miss. The Black Panther Party was for real and brought a new 'vibration' into the sociopolitical landscape of American cultural life.

In many other popular movies and television presentations this African American imagery is used in subtle ways to embody wisdom and healing, especially in this person as they age, a certain unexpected kindness, a vague familial compassion and certain 'magical Negro' qualities (Gonzalez,2001; Brookheiser,2001;

Parker, 2019). There are many examples of this phenomenon. It is a bit of a trope and quite controversial in the arts and entertainment industry. It has many of its roots in Blacks being seen as self-sacrificing as in those idealized conceptions of the loyal servant in antebellum days and later on as house servants to wealthy white people. But it may touch on deeper currents that go far beyond this. In the film *The Legend of Bagger Vance (2000)*, the Black character plays the Lord Krishna to the white reluctant warrior/hero Arjun to help him find his courage and rise to the occasion in the champion golf game; the film *O Brother, Where Art Thou (2000)* has a blind Black older rail line conductor as a seer seeing into the future and helping three escaping white prisoners on their Odyssey through evil lands. *The Green Mile* (1999) has a huge Black jailed prisoner, innocent we presume, healing two white people of seemingly incurable diseases by use of special powers. All of these connected the person to the larger spiritual world around them. It is my own belief that this may touch on complex reasons why African Americans in the United States are so intimately associated in literature and the arts with the moral and spiritual progress of America in general and whites in particular and partially serves as a sign and symbol of her greater unfoldment.

This *interiorization* of the African American in the United States also has had a paradoxical *exteriorization* and peculiar unifying effect. After all the presence of African Americans in a sense have 'helped' the diverse tribes of Europeans who came to the United States, be they German-Americans, French Americans, Irish Americans, Italian Americans, Greek, Swedish etc etc Americans, become, through the process of the 'melting pot' and group racism, largely drop their hypens and ethnic markings, intermarry with each other over the generations and simply meld into becoming 'white' Americans. This is by way of contrast to being 'Black' Americans. In their countries of origin they were Germans or Italians or Irish etc, but not 'white' people. That melding identity arose only on American shores and has been with us for centuries now. So historically 'Black vs. White' racism has been crucial to the development of a 'unified' United States. This is both a psychological symbolic process and a societal process measured by how we perceive 'the other's body' through color and behavior.

It is also true in this context that the African American body, especially the younger male body, is often ambivalently perceived as something innately powerful, as in certain contact sports areas, but also a symbol of implicit danger and a threat to 'white' America., prompting, often with little objective data, an unconscious rage and terror reaction that leads to official and unofficial overreaction and mayhem. It is currently an almost daily occurrence of, usually white male police and 'watchful citizens', opening up unprovoked and excessive violence on this symbolic object and stimulus to the collective American unconscious. Usually this occurs with few legal consequences and a wide pattern of acquittals even if it reaches a jury. The usual defense is something akin to 'I feared for my life' etc., which the white jury seems to emotionally 'understand'. This goes back centuries in the United States and is rooted in the African American body as exploited economic property, antebellum fears of slave rebellions and the later brief era of reconstruction in the south when whites needed to feel 'safe' and African Americans, especially the African American body, needed to be 'controlled' at all costs (Bynum, 2021). These fears are as old as the republic and their dynamics as close as the next national election.

In a real sense the African American presence and collective persona in the United States at least functions as a sort of moral and spiritual barometer for how American civilization treats 'the other' and serves at times as a literal measure of its ethical and behavioral progress in both world history and perhaps the spiritual evolution of our species. Again, this is systemic, collective and ancestral in nature and function.

Just in this narrow reference to the ancestral realm, finding one's 'roots' is now a wide socio- cultural phenomenon touched off in the USA by the book *Roots* authored by the African American writer Alex Haley. Decades later millions of the America people now send samples of their DNA in salvia or blood samples to businesses and corporations to search out their genetic roots for a fee. One of the most popular television programs today is called 'Finding Your Roots' hosted by the African American scholar and Harvard Professor Henry Louis Gates, an obvious esteemed elder on the American stage.

This is all in the exploration of the *family unconscious* (Bynum, 2005), a familial unconscious which at its deepest level is an *African Unconscious* in all humans, meaning *everyone's* African ancestry since all human branching's are, genetically and anthropologically speaking, from an original African template. In psychotherapy the early practice of role playing out one's inner conflicts, which often stem from familial dynamics and our actual family members, along with other deep role influences in the patient's life, was initiated by Jacob L. Moreno. Moreno meant 'brown hair or brown headed' but later in the former Spanish slave holding worldwidely came to be a derogatory name for Blacks. These are all ancestral influences and part of eldership.

What we are suggesting from this mythogenetic perspective is that on some intimate level, African American culture and its lived 'presence' has served to keep alive the feeling and memory of this inner luminosity and life-feeling in American culture ad civilization. Symbolically, and psychospiritually speaking, the music and cinemographic imagery we spoke of above are only two instances of this larger presence. It is also true in eldership and is in continuity with this lineage from the ancient past.

You see these elders and teachers in ancient times where believed to be the ones who had, in even pre-dynastic times, through their own discipline and commitment, awakened and raised the *djed* column in their own spines, unfolded the *ureaus* serpent, led it up to the crown of the head and liberated themselves, thereby communing directly with the gods of wisdom and the wider cosmos while passing this knowledge on to new generations (Bynum, 2012; King, 1990). These elders used many different religious labels over time to identify themselves, but underneath the meaning was the same. Invariably these elders returned from these journeys with narratives of phenomena beyond conventional perception. Often they depicted our known universe as a dense and smaller subset of a much vaster cosmos populated by beings and forms of intelligence that from time to time did commune with human intelligence. This universe, this *omniverse*, was essentially conscious and alive with life forms as diverse as could be imagined.

Resonance, Neuroscience and the Lineages of Spiritual Practice

It is useful to remember in our journey through this book that the dark neuromelanin of the brain's surface and areas of its folds, i.e., the so called gray matter covering the surface of both cerebral hemispheres, extends down and out the mid-brain limbic system into the center or core of the spinal line to near the base of the spine. This dark neuromelanin line again is *living* neural or brain matter. Its increasing presence in the mammalian line in our carbon-based life forms on this planet is highly and tightly correlated with increasing intelligence as we know it in evolution as pointed out in earlier chapters of this book. It has been an active process on our planet for millions of years and appears to be deeply implicated in the true *Roots of Transcendence (Bynum, 2006)* for our hominid line and playing itself out in our various forms of personality formation. To many of us it appears to be an intelligent and directed process, to exhibit a true pull toward the greater or *teleos* in our development. Of all the creatures of earth, those with the highest concentrations of neuromelanin in their brains and spinal lines are the mammals; of the mammals those with the highest are the great apes and primates; of the primates in reaches an apex in the chimpanzees. The only one with an even higher concentration is our line, homo sapiens sapiens, thinking man.

No, neuromelanin and its derivatives are *not* consciousness. However partially because it is inherently light responsive itself it seems to be our closest connection, to manifest our most subtle biological affinity for light itself. This neuromelanin, much like the so called *dark energy* associated with the literal *expansion* of spacetime itself in cosmology, appears to be associated with the *expansion* of consciousness in our evolutionary line. It is like the artistically rendered spice 'melange' in the science fiction *Dune* novels of Frank Herbert, except here it is not an artistic literary device but an actual fact of biological nature.

In this context it is interesting to note that there is an extracellular thread-like structure suspended in the CSF (cerebrospinal fluid) of vertebrates called the Reissner fiber that reaches from the brain throughout the central canal of the spinal cord. It helps in the development of the spine from the early neural crest days in embryogenesis onwards and if not there or damaged in some fashion contributes to various spinal pathologies, including scoliosis (Troutwine ,2020; Cantaut-Belarif,2018). It literally keeps the spine straight and parallels the flow of living energy through the spinal line. It does not appear to be the neuromelanin nerve current itself but does seem to act in close concert with it.

This entire process appears to us to be a neurobiological corollary of evolutionary intelligence and of the *djed* column or line. This *djed* column mentioned earlier was similar physically and symbolically to the Egyptian ankh or symbol of the life force sleeping in all human beings, literally *the pillar holding up the structure*, the force of unfolding evolution itself. Millions of the world's yoga practitioners of various disciplines implicitly acknowledge this daily without looking to deeply into its implications.

When, through discipline and meditative or spiritual practices of diverse kinds, this neuromelanin line within the spinal line, specific regions within the brain core and the surface of the cerebral cortex, is induced to '*vibrate* 'or *resonate*. Recall again the African American mythogenesis of *vibrainium* mentioned above. That was just an artistically rendered movie, but here actually does arise the possible aspects of what is termed biological superconductivity, a quantum mechanical process, which when awakened provides for rather significant increases in efficiency and bio-electric conductivity within the organism. We spoke of this earlier when we mentioned the brain and its *piezoelectric effects during coordinated movement and vibration.* Practitioners of certain forms of meditation report a clearer consciousness, finer focus of attention and expanded awareness on all levels of their existence.

On a deeper level this in all likelihood accounts for the innumerable reports of *perceptions of internal luminosity* in these elders and practitioners throughout the millennia across many disciplines and cultural contexts. This arises in various disciplines of classical meditation and in states of anomalous energy arousal where the body is seemingly frozen or suddenly paralyzed by light from an external source for which we have no widely accepted explanation within our current scientific paradigm. Scientists in our near future no doubt will take the exploration of resonance, neuromelanin and the spinal line in the unfoldment of consciousness into new and deeper directions, especially in healing and meditation and perhaps in nonlocal forms of connectivity in surprising ways throughout our wider solar ecology. This *nonlocal connectivity* is most likely attributable to resonance, which may provide this when spacetime itself is subsumed in a higher dimensional process.

The physics of resonance appears to underlie all three present day understandings of matter, to be able to interpenetrate and weave through linear space and time and in certain ways to be actually senior to or deeper than space and time. Conceptualizing what we classically identify as the 'soul' as a form of this deeper resonance provides a useful way to understand the notion of life and lifetimes purportedly observed in certain meditative disciplines and ritualistic conditions fluctuating around and through this noumena. We are referring to the initiation rituals within the great pyramids when the initiate perceived themselves as an extended identity through multiple lifetimes during the process of awakening. The fact that certain shafts in the great pyramid were designed to conduct the resonance of particular stars and constellations during the great year met with the internal acoustics of the inner chambers no doubt would have combined to open the student to the higher mysteries. We sense that the particular properties of neuromelanin in the brain and nervous system, specifically the capacity to transduce energy from one form to another, i.e., light or resonance to sound or acoustics, facilitated this underlying intimate process.

Also light itself seems to underlie this nervous system capacity for bioluminescence and consciousness. So for these reasons in the current scientific age our neuromelanin hypothesis is a leading candidate for the intermediary of these dynamics and phenomena and has a certain 'neatness of fit' for what we have been describing.

In a curious twist of how the collective unconscious works, this notion and literally the actual word for the *djed* column, represented in the hieroglyphs again as a column or pillar of stability for thousands of years (Faulkner, 2007; King, 2001), has migrated, through linguistic drift, over to the similar word and concept today as *Jedhi*. It is seen in the worldwide positive collective response to the Star Wars cinema-mythology and their supposed *Jedhi* masters, i.e., *Djedhi* masters. We say *mythology* and not science fiction due its impact and meaning in collective cultural terms. This term also has biblical roots and suggestions in *Jeddah* or *Jedidiah*, i.e., being beloved of the god Jehovah and in divine communion and favor.

For thousands of years these rites were practiced not only in the three great pyramids of the Giza plateau we have mentioned, but also by countless other initiates in other pyramids still buried beneath the sands and only recently revealed by modern satellite technology (Tucker, 2016). Think about that for a minute, a whole culture and psycho-mythology for countless generations associated with disciplines devoted to the exploration of a consciousness we are only beginning to rediscover in our day. What narratives could be told and rediscovered from this era? Modern literature and even internet websites are full of these accounts by both ancient and modern chroniclers (Prada, 2017; El-Amaan, 2017; Haich, 2000). Music recorded in the Great pyramid reflects how its unique acoustics are conducive to entering these states of consciousness (Horn, 1997). Much of this hinged on dreaming, controlled dreaming, ritual and related methods. These in later centuries in somewhat watered down versions after the fall of Kemet spread around the Mediterranean basin in the famous sleep temples of the Greeks and Romans and throughout the Black diaspora.

Some derivatives of these Kemetic disciplines and the science of consciousness embedded in them are manifest today in North and South American as well as Caribbean practices of IFA and the different disciplines of the Orisha with their various religious branchings, e.g.,Candomble, Santeria, Lecume etc . In the United States certain cities with larger African American populations, e.g. NYC, New Orleans, Chicago, Los Angeles etc, have diverse communities where variations and currents of these practices thrive and intermingle with each other. Today they are often rich in ritual and religious symbolism but sometimes thinner in the actual somatic transformative disciplines and the deeper trance work.

It has not been that much of a problem until modern times in African American traditions that when elders became older and older they could feel their 'god' or spirit was in a great tree near them in the countryside or that ghosts and family spirits visited them nightly for many reasons and they could directly commune with them. There was no implied 'dementia' simply because of this. It was assumed that the elder was slowly detaching, transitioning to the ancestral realm or another level of vibration. The feeling of life was the crucial thing. In those days they had guides into this new realm, something sadly missing today. All of this is what we were alluding to earlier when we suggested that African American culture kept alive the memory of this inherent inner luminosity and life feeling in Americans who were immersing themselves in our materially dominated age and in some ways has stood as the guardian of this spiritual dimension in American civilization. It keeps alive the feeling of spirit and in no way denies the reality of medicine or science.

Much of this knowledge lives like a hazy dream in the unconscious just below the surface of millions of African Americans and others scattered throughout the diaspora. But they are made wary of it by the fearful images of being seen as "unscientific", of studying 'Black magic', of succumbing to the lore of 'primitivism', the fear intoxicated visions of juju/voodoo and all the other psycho-cultural images peoples of direct African descent, or anyone else for that matter, risks being labeled if they explore to far off the beaten track.

Eldership, Divination and Your Everyday Laptop Computer

From what has preceded this section so far you might think all this has little to do with the technological world we currently inhabit. Nothing could be further from the truth. As mentioned earlier these elders and priests of ancient Kemetic Egypt have a direct connection to our society today not only in the depth psychology we have explored but also in the operation of your phone, your laptop and many other digitally based technological devices.

During their practices in the pyramids and other sacred sites these elders and priests practiced divination, as did and does much of traditional Africa today. For the Egyptians the sacred system they used was a base 2 whereas the Kabbalah 's Ayin Sof is a base 10 emanated by 10 Sefirot. The Christian and Hindu systems used the decimal system of base 10. This is important because the use of base 2 places an emphasis on doubling, which was widespread in Africa, i.e, the sacredness of twins, doubles in the spirit world etc. Both Diodorus Siculus and Oblian noted that the ancient Egyptian priests and elders "employed an image of truth cut in halves" (Kautzsch,1912). There is also the vision of Thoth/Hermes of "as above, so below, as within, so without". There is ample archaeological evidence linking this use of doubling in counting systems throughout sub-Saharan Africa (Zaslavsky (1991), Bamana sand divination, which has been around for centuries, being only one example of this practice of divination or geomancy in the earth. A very similar system of divination in Africa between the Ifa and Fa has been noted by others (Trautman,1939).

With the Arabian incursions into Africa in the 7th century their scholars and scientists noted and absorbed some of this into their own systems. These Islamic scholars began documenting it by at least the 9th century according to the Jewish scholar Aran ben Joseph even though it had been ubiquitous in Africa for millennia before this era. The Arab scholars had already learned the Egyptian mathematics of the 'gebra' which became al-gebra or algebra. Then in the 12th century Hugo Santalia brought this code of divination from Islamic mystics to Spain where it entered the alchemy community through geomancy or earth divination. This geomancy was usually 4 bits of random dashed lines with 1 or 2 strokes in the sand that are repeated, a divination system seen in many parts of Africa (Eglash,1995). This geomancy chart was shown to King Richard II in 1390 and later mentioned in the 1600's by German mathematician Leibniz in his dissertation "De Combinatoria" where he also spoke of geomancy and introduced the binary code. Leibniz replaced dashed lines for 0's and 1's (Eglash,1991).

In the mid 1800's the mathematician George Boole took Leibniz's binary code and created Boolian algebra. In 1877 the mathematician Georg Cantor took these lines and created the Cantor set by making these lines recursive. Boolian algebra laid the foundation for the digital computer. Eventually von Neumann took Boolian algebra and created the modern computer system that underlies your laptop, smart phone and innumerable other digital devices. So in many very direct ways the ancient system of divination is rooted in an implicit binary system underlying the divination systems of ancient Egypt and others widespread areas of Africa. Such is how human ingenuity and cultural cooperation over time lays down the bricks on the road that all human progress walks on today. The relationship between the divination systems of our elders in geomancy, dreams, mathematics and so-called mysticism is a long, multifaceted and complex interplay of people, peoples and events. We doubt very much that anyone denies how the system underlying the computer does not in its own way provide from the divination of data in the modern world. And yet we still fear this intertwining of dreams, psychology and science.

Eldership, Dreaming and the Descent of the Noosphere.

These fears and all our work on dreams and symbolism of course has ancient roots. Our species Homo Sapiens Sapiens in its present anatomically modern form has been around for anywhere from a conservative 150000 to a more likely 200000+ years. All this time as modern humans we have been thinking, dreaming and roaming the earth in search of food, meaning and the gods. We have loved, wared, mated and wondered with each other for untold millennia. We have cross fertilized innumerable times and various ethnic diversifications within our species have come and gone as have cultures and civilizations. Our DNA, anthropological bone excavations and serological markers have given unmistakable testimony of this to modern science.

Undoubtedly in our early period we were fully emersed in 'nature' along with the other animals of the forests and savannahs. But by the time we began to emerge from our Paleolithic or 'stone age' we had slowly begun to detach ourselves from total immersion in the so called natural world and actually reflect upon it. No doubt we began to notice our dreaming process and consciously began to see it as part of and a reflection in some way of our reality and thought and the dynamics of the world. Our great hunters and elders had appeared in the tribes and social structures who helped interpret and manage the world for us. We began to reflect enough upon nature to intuit that something subtle and unseen in us existed that went beyond us and so began to bury our dead. This began sporadically at first but then it became a marker of our species as it did earlier in our Neanderthal cousins from time to time. Our thinking processes around the globe where ever we traveled to and dwelled created what some have referred to as a *noosphere*, a kind of mental and psychospiritual envelop around the earth for thousands of years (de Chardin, 2008; Vernadsky,2014). It became the seed form of what would later evolve into our collective unconscious. Arising from the geosphere and then the biosphere, it set the stage for a truly expansive interplanetary form of communication for the future of man. Today we have constructed vast electromagnetic and information sharing systems around the earth that technologically express this development. It is worth noting here that since we were until somewhere around 25000 to 35000 years ago all African in phenotype from our 200000+ year history and millions of years of pre-history in Africa, its contours and functioning are essentially rooted in **Our African Unconscious** (2021).

We were behaviorally modern humans. Then somewhere in the middle of our upper paleolithic age, perhaps between 50000 and 40000 years ago, we suddenly gave rise to an abundance of symbolic activity regardless of wherever we were distributed on the earth. Dreams, elaborate rituals, magical beliefs and perceptions, symbolism and all their expressions in artifacts, cave paintings, petroglyphs and engravings on bones and ivory proliferated. The noosphere, surrounding and infusing the earth for thousands of years before this then seems to have *descended* into our dreams, into our spiritual and ritualistic life. The elder archetype became even more essential in interpreting the world. The world above and around us descended and met with that subtle living light within us to integrate into a kind of Kim Wirian synthesis. Religion and philosophy had taken a quantum leap in our collective experience. The masters of the meanings and rituals were the elders and the ancestors. This has continued into our own day.

Dreams and Eldership Today

Of course, some things about dreaming do not change simply because of culture or the times or the Age and aging. Women today still dream in more bright colors, know who their intimate sexual partners are more and tend to be more relational in general. Men still tend to have much less color in dreams and still after years have more anonymous sexual partners. Men dream about money more, women more about their children. Both have spiritual lives deeply implicated in the dreaming process.

However as we age and become elders we tend to dream less about our careers and past glories and more about of our spouses, our families, those we love and our grandchildren. We are given the freedom or perhaps take the license to dream more about those in our lives and family members who came before us, helped and healed us in intimate ways, and taught or initiated us into the deeper interior meaning of our personal lives. We acquire, if we are open to it, a deepening sense of how our lives are subtly and intimately interconnected with each other. These connections have and do mold our emotional lives and, given the subtle plasticity of the brain, molded the contours of the brain and mind itself. They have literally *in-form-mated* us.

An essential part of eldership in any Age is reminiscing, summing up one's life and contribution to family, society, sometimes even to civilization. We experience either ego integrity or despair at this stage. Memories are stored and recalled not so much in a linear fashion, but more in accordance with profound influences on us and their energetic qualities. In some instances we are charged with recording the collective memory and lineages of our society as *griots*.

Unfortunately, the age appropriate process of a slow withdrawal from society, growth into wisdom and increasing other dimensional perceptions of the world has atrophied, lacking appropriate role models and disciplines in the current era to navigate this stage. Much to often today eldership has degenerated into a kind of cataloging of dementias and Western society in particular has largely medicalized this whole era. Again we have thrown the baby out with the bath water. The dream has almost died out as a way of understanding and moving from one state of consciousness to another. It is a problem that will only get worse until we shift our orientation.

Obviously, we are not saying all dementias have no clinical basis, not at all. Rather that, like 'spiritual emergencies' in psychiatry and psychology, symptoms can be medical-clinical in nature but also have mixed noetic features. The American Psychological Association diagnostically recognizes this phenomenon in some contexts such as in conflicts around religious perceptions.

So by no means does this necessarily imply decreasing the use of certain medications and clinical procedures in palliative care, only seeking a restoration of the balance between compassion and respect for the deeper dimensions of consciousness and an open awareness of the reality we are all passing through, that has in-formed us and we have known for millennia. Consciousness is the ground and interwoven everywhere. The cosmos above us and the luminous cosmos below the quantum begin to be reunited in these situations, especially with elders and their dreams.

Indeed there appear to be vast spaces in and below the quantum our awakened elders and ancestors explored on the banks of the Nile and Ganges in states of deep meditation and visionary contemplation that revealed worlds as real to their consciousness at that level as are the objects to us at our level. At that level 'far' and 'close', 'huge' and 'tiny' fade into relative categories through which we perceive and organize reality. Their vibratory confluences and resultant expressions were the basis for empirical methodologies of observation long forgotten or as yet unrealized by modern science. This includes the embodied life of the soul. Given the quantum potential at both the very inception or entry into earthly life in the womb and its end in terms of luminosity and the nonlocality of energy and consciousness in the NDE and beyond, it is at present unclear as to the range and distance consciousness and even the ecology of souls encountered is in a cosmos spanning these episodes (Bynum, 2021). The range of the disembodied soul is a mystery that we might well *not* put limits on at this juncture.

Before this time our dreams and the personas in our dreams hinted at a reality populated by personhood and consciousness that was seemingly all pervasive, pathways crossing, lifetimes enfolding each other. Now this becomes fully unfurled in moments and flashes of radical insight when space-time and distance are collapsed and a trajectory beyond words and conception arises.

In a future science we may yet see a luminous intersection of our popular family genetic testing connections that we mentioned earlier and are currently so enchanted by with a more coherent understanding of our dreamlife recollections in order to trace our lineage as traveling souls through numina and phenomena that as of yet makes little sense to us in the current era. Our different dreamscapes may be quite interconnected with each other in luminous ways, especially when we consider how our lives are intimately connected with others in the same life envelope of relations, years and space-time episodes (Bynum, 2017).

Our whole human panorama of experiences from birth to death may appear as a kind of prenatal womb or staging area for a further development beyond the current human form itself. This implies an evolution of consciousness in which we are fully *translated* into beings of light beyond the *bardos* briefly encountered in the classical NDE or near death experience and the OBEs or out of body experiences of the pyramid initiation rites. The luminous being that emerges in innumerable clinical reports of the NDE experience, regardless of religious, or cultural, or historical era or philosophical perspective, bears witness to how inextricably in-formed and interconnected we are with so many significant and minor others during our earthly embodied episode. Just as our dreams are interconnected, so are our other embodied lives. Only those in complete denial and addicted to early 19 and 20th century forms of logical positivist rationalism and pre-quantum understandings of solidity and matter dismiss this recurrent testimony of human experience. Here may be a glimpse into the farther reaches of evolution itself embedded in the seed of human consciousness that will eventually be unfolded.

In the meantime we must acknowledge here that our dreams of problem solving do certainly take up much of our dreaming life no matter what our age or place in the epigenetic stage of the family lifecycle. As we age and reminisce we dream more openly of death and transformation with less and less fear. Those who have died before us and entered into what many traditional African societies refer to as the great *Zamani* (Mbiti, 1990)or zone of the deceased who still have some vibrational contact with the living on some level of consciousness, are encountered with less fear. In the past they would have been approached with hesitation as members of 'the living dead" with several folds or levels within this realm of consciousness. In dreams and the NDE they now manifest to us more easily and often with great delight. It is the realm of the ancestors that has always been there and a hint as to the kind of beings we really are enfolded here in dense matter and our place in 'the great chain of Being'.

This is the door that was open to us in the remote past and still survives in some indigenous societies. Someday soon this realm of human experience will be opened to us clinically in psychology and psychiatry as the farther reaches of our extended familial unconscious. Elders sometimes openly commune with the energy of this spirit world to the disquiet or amusement of family members. Those who are leaving but who are remembering, wide awake and at peace with themselves see the full panorama of life, becoming a "chanter of pains and joys, uniter of here and hereafter".

Yes, as we said, to many of us think of all of this as only 'dementia' and to often do not see its psychospiritual roots or potential. Talking of this can be unsettling to others around us, especially when we are ill, but we are often at a resolution point after a time of emotional and mental integration. If the door is open to discussing this in any of its permutations within the life and family context of the person, this is the time to do it. It is also when eldership is most valuable. The spiral of life is slowly spinning and a wormhole into another dimension is opening. Someday, when we are no longer the noisy silly beings of today, we will travel back and forth freely between these realms of light and realize it is an ancient science.

In the meantime, whether they come from the wind-swept plateaus of ancient Egypt or the dense forests of West Africa or the bustling parishes of the Americas spread far and wide, when some elders who have passed over visit us in the emergency room or in our dreams, they are opening us to the next chapter in our human experience and perhaps more, perhaps something that the noosphere can only point toward. Spirituality, once dismissed in adolescence, often takes on a deeper more profound resonance. It is the final stage of growth to which our dreams bear witness.

REFERENCES

Akiwowo, A. (1980). *"Ajobi and Ajogbe: Variations on the theme of socialization"*. Inaugural lecture, Ile-Ife, Nigeria: University of Ife Press. June 10th.

Ben-Jochannan, Y. A. A., (1970). *African Origins of the Major 'Western Religions'*. Black Classic Press, Baltimore: MD

Barr, F.E. (1983). "Melanin: The organizing molecule". *Medical Hypotheses*. 11, 3-4.

Bauval, R., & Brophy, T.,(2011*). Black Genesis: The Prehistoric Origins of Ancient Egypt*. Bear and Company, Rochester, VT.

Bernal, M. (1987). *Black Athena: The Afroasiatic Roots of Classical Civilization*. Rutgers University Press, Rutgers: NJ.

Bonwick, J. (1878). *Egyptian Belief and Modern Thought*. London, UK: C. Keegan Paul.

Brookhiser, R. (2001). "The Numinous Negro: His importance in our lives; why he is fading". *National Review*, August.

Brophy, T.G., (2002*). The Origin Map: Discovery of a Prehistoric, Megalithic, Astrophysical Map and Sculpture of the Universe*. NY: Writers Club Press.

Browder, A. T.(2004). *EGYPT on the Potomac*. Washington, D.D.: IKG Publishing.

Bynum, E.B.(2017). *The Dreamlife of Families*. Rochester, VT: Inner Traditions & Bear Company.

Bynum, E.B. (2012). *DARK LIGHT CONSCIOUSNESS: Melanin, Serpent Power and the Luminous Matrix of Reality*. Inner Traditions and Bear Company, Rochester, VT.

Bynum E.B. (2021). *Our African Unconscious: The Black Origins of Mysticism and Psychology*. Inner Traditions and Bear Company,

Rochester, Vermont. (Expanded edition) (The African Unconscious, NYC: NY. Cosimo Books, 2005.)

Bynum, E.B.,(2006).*The Roots of Transcendence*. NYC:NY. Cosimo Books, formally published as *Transcending Psychoneurotic Distortions*. Haworth Books, 1994.

Bynum, E.B. (2005). *The Family Unconscious: An Invisible Bond*. Cosimo Books. NYC: NY.

Cantaut-Belarif, Y, et al, (2018). The Reissner Fiber in the Cerebrospinal fluid controls morphogenesis of the body axis. *Current Biology*. August 6.28 (15),2479-2486.

Chandler, W.B. (1999). *Ancient Future: The Teachings and Prophetic Wisdom of the Seven Hermetic Laws of Ancient Egypt*. Black Classic Press. Baltimore: MD.

Cerami, C.A., (2002*). "The Dogon Ancestor". In Benjamin Banneker: Surveyor, Astronomer, Publisher, Patriot*. Appendix I. John Wiley and Sons, New York: NY.

Collins, A., and Little, G.L.,(2022). *Origins of the Gods: Qesem Cave, Skinwalkers, and Contact with Transdimensional Intelligences*. Inner Traditions & Bear Company. Rochester:VT.

Creighton, S., and Osborn, G. (2012). *The Giza Prophecy: The Orion and the Secret Teachings of the Pyramids*. Bear and Company, Rochester: VT., 217-219.

Currivan, J.,(2017). *The Cosmic Hologram: In-formation at the Center of Creation*. Inner Traditions. Rochester, VT.

De Chardin, P.T. (2008). *Phenomenon of Man*. Harper Collins, NYC;NY (1948).

Diop, C. A., (1991). *Civilization or Barbarism: An Authentic Anthropology*. Lawrence Hill Books, Brooklyn: NY.

Diop, C.A., (1974). *The African Origin of Civilization: Myth or Reality*. Lawrence Hill Books, Brooklyn: NY

Edmonston, W.E. (1986). *The Induction of Hypnosis*. Chapter 1, "The Ancients". NYC; NY. John Wiley & Sons. 1-25.

Eglash, R. (1991). "Africa in the Origins of Binary Code", in *African Divination Systems*, by P.M. Peek (ed).,Indiana University Press, Bloomington, IN. 112-132.

Eglash, R. (1995). "African Influences in Cybernetics", in C.H. Gray (ed) *The Cyborg Handbook*.

El-Amaan, (2017). The Life of an Egyptian Initiate. *www.crystalinks.com/initiationegypt2htmi*.

Faulkner, R.O.(2007). *The Ancient Egyptian Pyramid Texts*.Digireads.com Publishing. Stilwell, KS.

Finch, C.S,(1998). *The Star of Deep Beginnings: The Genesis of African Science and Technology*, Decatur GA,: Khenti, Inc Publishers.

Gonzalez, S. (2001). "Director Spike Lee slams 'same old' black stereotype in today's films ". *Yale Bulletin & Calendar*. Yale University, New Haven: CT. 03-02.

Griaule, M., and Dieterlen, G. 1986. *The Pale Fox*. Chino, CA: Continnum Foundation.(originally published in French as *Le Renard Pale*, Paris, 1965)

Haich, E. (2000). *Initiation*. Santa Fe: NM. Aurora Press.

Hancock, G. and Bauval, R. (1996). *The Message of the Sphinx: A Quest for the Hidden Legacy of Mankind* . NY. Crown Publishers, Inc.

Horn, P. (1997). Inside the Great Pyramid. Amazon Music (re-issued)

Hourning, E. (1986). "The discovery of the unconscious in ancient Egypt". *Spring Publication: An Annual of Archetypal Psychology and Jungian Thought*.16-28.

Jackson, J.G., (1970). *Introduction to African Civilizations*. Citadel Press, Secaucus: NJ.

King, R.D. (2001). *African Origin of Biological Psychiatry*. Germantown: TN. Seymour-Smith Publishers. (1990).

Krishna, G.,(1978). *Yoga: A Vision of its Future*. Kundalini Research and Publication Trust, New Delhi, India.

Kautzsch, T. (1912). "Urim", Encylopedia of Religious Knowledge.

Mbiti, J.S.,(1990). *African Religions and Philosophy*. Heinemann International. Oxford: UK.

Mills, A et al, (1994). "Replication studies of cases of suggestive of reincarnation by three independent investigators", Journal of the American society for Psychical research.,88, July

Morakinyo, O.(1983). "The Yoruba Ayanmo Myth and Mental Health Care in West Africa". *Journal of Cultural Ideas*. December, 1 (1)61-92.

Ozaniec, N. (2022). *Becoming a Garment of Isis: A Nine Stage Initiatory Spirituality*. Rochester,VT: Inner Traditions & Bear Company.

Parker, M. (2019). *Magical Negro*. Tin House Books, Portland, Oregon & Brooklyn,: NY.

Prada, L. (2017). An Egyptian Initiation. (Brother Veritas website). *www.luisprada.com*.

Shroder, T. 1999. *Old Souls*. Simon and Schuster. New York: NY

Schwaller de Lubicz, R.A. (1998).*The Temple in Man*. Rochester, VT: Inner Traditions and Bear Co.

Schwaller de Lubicz, R.A. (1961). *A Sacred Science: The King in Pharaonic Theocracy*, Rochester, VT: Inner Traditions and Bear Co.

Stevenson, I. (1995). *Twenty Cases Suggestive of Reincarnation*. Charlottesville; University Press of Virginia.

Stevenson, I. (1997). *Reincarnation and Biology, Vol. 1: Birthmarks and Vol.2; Birth Defects and Other Anomolies*. Westport CT: Praeger.

Trautman, R. (1939). "La divination a la Cote des Esclaves et a Madagascar, Le Vodou, le Fa, le Sikidy".*Memories de I'institut Francais d'Afrique Noire.*

Troutwine, B.R. et al, (2020). The Reissner Fiber is highly dynamic in vivo and controls morphogenesis of the spine. *Current Biology.* June 22, 30 (12), 2353-2362.

Tucker, A. (2016) "Space archaeologist Sarah Parcak uses satellites to uncover ancient Egyptian ruins". *Smithsonian Magazine,* December.

Van Sertima, I.,(1976). *They Came Before Columbus: The African Presence in Ancient America.* NY: Random House.

Vernadsky, V.I., Rouillard, M.K., and Ross. J.A. 2014. !50 Years of Vernadsky: The Noosphere, Vol 2. Create Space Publishers (1938).

West, J.A. (1993). *Serpent in the Sky: The High Wisdom of Ancient Egypt.* Wheaton, Il : Theosophical Publishing House. pp 44-45.

Whitman, W.(1964). *Out of the Cradle Endlessly Walking* .Leaves of Grass. New York: NY. Signet Classic. 1882.

Zaslavsky, C. (1991). Africa Counts: Number and Pattern in African Cultures (1st edition 1973). Brooklyn ,NY: Lawrence Hill Books.

EPILOGUE

ON. I give you the Eye of Heru, because of which the Gods were merciful. ON. I give you the Eye of Heru; betake yourself to it. ON. I give you the lesser Eye of Heru, of which Seth ate. ON. I give you the Eye of Heru, with which the mouth is opened. The pupil which is in the Eye of Heru, eat it. ON. I give you the Eye of Heru, and you will not be ill.

-Utterance 935, Coffin Texts, 2100-1675 B.C.E. Kemetic Middle Kingdom (Faulkner, 1978).

POSTSCRIPT

Data is not knowledge and knowledge is not wisdom but when ancient wisdom is married to new knowledge it may suggest a radically transcendent vision of reality. From the introduction we sought to place this study within the context of humankind's ongoing scientific exploration of itself and its place in the cosmos. The more our species has explored the world within itself, the more it has seen itself reflected in the world outside. "As above, so below, as within, so without", the principal revelation of Tehuti and the core of the Hermetic corpus of ancient writings became the root metaphor of a scientific and spiritual intuition that in the Ages that followed has given birth to innumerable discoveries, inventions and intimate understandings of ourselves. Like mathematics it yokes us to the world process and its most luminous permutations.

Beginning in the first chapter with Professor Moore, we tried to harness the new knowledge of embryogenesis or life and development in the womb to focus on how evolution's drama of increasing complexity was intimately involved in the dynamics of melanin and neuromelanin. Melanin's capacity to interact with light and transmute it to higher levels of order and complexity, to transform it from one state of energy to another, was seen as a crucial parallel to evolution itself. Indeed melanin and neuromelanin in particular was no longer seen as a mere biochemical "waste product" of the nervous system but rather an intimate player in the drama of life's mysterious unfoldment into more rarified expressions of intelligence and light. The fact that neuromelanin appears to absorb light and to increase in density and amount as we progress or "ascend" up the evolutionary ladder only deepened this sense of an intimate embrace with the dance of evolution. But melanin was not restricted to the brain or even the human or mammalian body. Professor Brown's chapter elaborated on the different types of melanins found in the wider environment, including the soil, the water, the air, indeed in the wider solar system itself. The startling fact that it is also localized in strategic locations within the inner most recesses of the brain only reinforced the sense of importance of this neglected area of science. Perhaps it is because melanins are found from the reaches of the solar system to the inner sanctum of the brain core, that the widest "outer" in some physical and energetic sense reflects the deepest "inner", denies us a clear boundary between these realms and is a confession that in some still mysterious way they reflect each other. Science can never be completely divorced from metaphysics any more than energy can be from information and yet the two are not identical.

In our search for the meaning of life and our place amid the stars, we as a species have turned our attention to the seat of our own consciousness, the brain. Like neuroanatomy itself in the early days of Kemetic Egyptian exploration of the body in the process of mummification much was discovered about mental functioning based on neuroanatomical structures. Melanin and neuromelanin exploration in neuroscience has furthered this understanding by noting the strategic location of neuromelanin in these brain structures, especially beyond the surface cerebral and into the deeper limbic and other sub cortical structures. By adding to it the influence and interaction of biochemistry and evolutionary neurobiology to the dynamics of light the study of neuromelanin has implicated the subtle loom of consciousness itself. It has an affinity to light. Light remains our most fundamental mirror and intuition of matter itself.

By the mid point of this book, we turned out attention to consciousness itself and its involvement with neuromelanin. Beyond the medical and physical involvement of neuromelanin with stress symptoms of psychosomatic disorders it was emphasized how the mid brain limbic system of neuromelanin foci interacted with higher cerebral structures in the modulation of emotions, feelings, images and memories. Emotionally powerful symbolic images were seen as particularly evocative.

Essentially, we suggested that in a very real way we consciously interact with neuromelanin through our emotional and somatic experiences. This includes not only our personally informed memories and experiences, but also our socially informed collective and deeper contemplative experiences born of our psychospiritual disciplines stretching back millennia through innumerable traditions. Realization in the contemplative disciplines leads to the dissolution of all experiential categories of internal and external space and opens into a luminous confluence of self, other and numinous processes in which space, time, matter and location are derivative notions. This was as true in antediluvian Kemet on the banks of the Nile before the raising of the pyramids as it was in ancient Tibet on down to the refined meditative disciplines of the post-modern age. Neuromelanin and its affinity to light and consciousness is an intimate aspect of our nature and is here reemerging into the context of contemporary science.

Tehuti, Hermes Tresmigistis, the inward systems of order in flowing permutation with the wider outside cycles of the cosmos, this is the ancient vision coming to rebirth in our times. Nowhere is this more in evidence than when Dr. King explores the stellar dynamics of the solar expanse and the intimate process of the neural net.

Neurocosmology is as old as the human mind itself and rooted in the same soil as its progenitors in Africa. It is indeed an irony of this age that with this being rediscovered the very peoples who initially discovered and formulated this principal of the world process, the peoples of northeast and west Africa, should find themselves in the economic and geopolitical fix that they do today.

There have been many rises, many falls, many forgotten triumphs and resurrections of the peoples of Africa. But how could it not be so, for Africa is the genetic and anthropological root of the human species with all its permutations and dramas and the deep brain core carries this primordial narrative through all its variations in the diverse peoples of the earth. Today, despite the daily drumbeat of negative news about poverty, disease, war, malnutrition, corruption, indeed every plague known to humanity attacking the continent and its people, Africa is again slowly ascendant. In recent centuries the fall of Africa again after another cyclical rise during its medieval ages lead to an unprecedented spread of African slavery and degradation. Dark skin became identified with a fallen state and took on near mythical proportions, spreading its racially infected ideology into the tacit assumptions of science, religion, politics and the intimate dynamics of social and family life. Dark skin became a sign of a lessened status, a fractional membership in the human family. The very study of anything dark-skinned or even associated with dark skin became taboo or else feared, denigrated and devalued.

A similar history occurred in Dravidian India when the Indo Aryans swept down from the north and through war executed over a hundred years essentially enslaved the dark-skinned inhabitants of the Indus valley. The British episode reinforced this notion. It is a curious irony that in both narratives the dark-skinned inhabitants had created and nurtured, centuries before the invaders, highly disciplined contemplative sciences that sought to raise and fill the inner sanctum of the human mind with a luminous reality that connected it with the stars and a reality beyond linguistic conceptualization. In the Indus valley it was called yoga, in Kemet it was a similar science. The Indo Aryans took the science to themselves and over time forgot its origins in the low caste dark-skinned peoples they came to despise. Early Islam, Christianity and Judaism all have their historical roots in the soil of dark-skinned peoples. This is no coincidence. It was no accident either that these two subjugated civilizations for thousands of years before their conquest were in contact with each other through the land and sea lane trade routes of the Indian ocean. What is crucial for our study here is that in both cases they discovered how to illuminate the inner darkness of every human mind with a kind of living light that was an intimate part of the body, the brain and the deep brain core. That brain core vibratory reality we suggest here is an activated and awakened neuromelanin nerve tract.

In the post modem Age identity politics is all to real. The taboo of embracing our collective dark-skinned African origins living in our very brain core is ironically similar to the taboo that prevents us from embracing the paradoxically dark luminosity this brain core and its surface confesses about who we really are and where we all come from. Both in many cases leads to the dissolution of a more dominant politically, culturally and even psycho religiously informed sense of self while offering instead an uncharted, boundless and terrifyingly radiant state of consciousness.

People often want a tan but also the perks of white privilege. People want to be religious but generally do not want to have a religious experience. Therein lies our postmodern dilemma and we are all dimly aware of it. We just don't quite know what to do about it either.

We are suggesting here that the study of melanin and neuromelanin may just offer us a way to collectively heal ourselves and transcend this dilemma. Melanin on the skin's surface in recent centuries in large areas of the world has tended to be used to divide us, dehumanize us, teach us to marginalize our humanity and contract our collective sense of an interwoven identity. Now neuromelanin, an anlagen of our shared origins, an echo of our collective identity rooted in the nerve work itself, suggests a pathway out of our spiraling nightmare of hate, recoil and rampages of revenge.

It is one of the living realities that unites us in a deep place and connects us through time and differentiation with all the higher life forms of this earth. It has been there since the beginning and in a real sense been part of the conductivity of our life force as we have ascended upward through the evolutionary arch. It has quickened our nervous activity, deepened our consciousness and made subtler our apprehension of the seen and unseen world. No it is not the sole axis of human development. Rather it is one of the fathomless currents that flow through the oceanic mystery of who and what we are.

Someday our future science will turn its attention to what occurs vibrationally when, through discipline and technique, the internal constellation of neuromelanin enters into a conscious resonate affinity with the solar ambience of melanin particles distributed throughout the solar expanse. Everything is luminous and alive moving through transformations and translations. When that occurs our ancient insight that we are luminous beings and our new knowledge based on science and technology will emerge onto a vast new plane of experience. At that time when collectively 'the eye is single and the body is full of light', our progeny living on the shores of distant planets will realize the vision worked out on the banks of the Nile dateless millennia ago.

GLOSSARY

Afferent- These are structures which conduct fluid or nerve impulses toward an organ or other structures, e.g., afferent neurons conduct impulses to the central nervous system.

Australopithecines- Any of several extinct bipedal pre-Homo primate ancestors from millions of years ago. They are known primarily from the Pleistocene fossil period in southern and southeast Africa.

Avian System- Studies or other research having to do or associated with bird and birdlike systems.

Blood-Brain Barrier- A system of barriers (membranes, etc.) that inhibit the passage of certain molecules from the blood into brain tissue and cerebrospinal fluid.

Catecholamine- Any of a group of amines, which includes epinephrine, norepinephrine, and dopamine, that are derived from tyrosine and have a hormonal function.

Cell Culture- The growing of cells invitro (glass), including the culture of single cells. In cell cultures, the cells are not organized into tissues (tissue culture).

Cytosol (perikaryon)- The cytoplasm around the nucleus in the cell body of a nerve cell. This is the liquid portion of the cell, in which is suspended other internal cell structures.

Dark Matter- The unseen so-called "cold dark matter" and/or energy that accounts for about 93 percent of the matter in the cosmos as presently known. It is detected by gravitational measures but is currently of unknown content.

Deafferentation-The decrease in neural impulses or input to specific regions of the brain, which results in a change in the usual function of that area and its parallel psychological process.

Efferent- Pertains to those structures that lead away from other organs or structures, such as the efferent arteriole of the kidney nephron.

Genius- The genie or developing angel, the second stage of psychospiritual human development in the Kemetic mystery school system. The second stage of development unifies both the left cortical "masculine" logical consciousness and the right cortical "feminine" emotional consciousness to produce a unified consciousness, with an expansion and elevation of passion, logic, elevated neurogenesis, and creativity.

Histological Study- A microscopic study of the minute structures of cells, tissues, etc., in relation to their functions. Nerve cells are studied using a preparation under a microscope.

Horus, or Heru- The ancient Kemetic Egyptian god of the sun and light as represented by the hawk-headed figure. Often associated with the life force ascending beyond death and associated with the sons/daughters of light. That third state of human development that results from conscious unity with the ancestral realm and the perfect fulfillment of one's core creative passion, mission, and purpose for being.

Morula- The spherical embryonic mass of blastomeres formed before complete blastulation.

Myelin Sheath- A fatty sheath-like covering of the axons of some nerve fibers. Nerve fibers that possess myelin sheaths are called myelinated nerve fibers, and those that do not are called unmyelinated nerve fibers. The myelin sheath protects the nerve fiber and transports impulses rapidly.

Neural Crest- An area bordering the neural tube that forms as an invagination of the ectoderm layer, from which cells migrate and later become specialized into melanocytes, ganglia, and many other cell types as they migrate to endocrine glands and other body structures.

Neural Tube- A tube formed from the neuroectoderm of the early embryo by the closure of the neural groove; it develops into the spinal cord and brain.

Neurocosmology- The study of the subtle interrelationships between the brain and the cosmos as expressed in metaphor, symbolism, and physical and /or energetic principles and correspondences.

Neuromelanin I-33 Tissue of Heru- The Black Ben-Stone of neural tissue in the brain and spinal column that transforms the experiences of living biological systems from the lower animals to the higher, then to the human, higher human, and eventually to the transcendent realms of experience and expression.

Neuron- Any of the cells of nerve tissue, consisting of a nucleated portion and cytoplasmic extensions.

Parietal Lobe- One of the four major sections, or lobes, of both hemispheres of the brain. It is interconnected with all other brain areas and is associated with sensory orientation and motor functions, along with other neural processes. The other lobes are the frontal, occipital, and temporal lobes. Between the cerebral cortex and the brainstem is another lobe, often referred to as the limbic lobe, through which information and emotional impulses flow.

Phylogenetic- The evolutionary development of animal and plant species. Their historical and cultural development.

Pluripotent Stem Cells- Usually refers to progenitors of blood cells. Cells from the neural crest that have the potential to become many different cell types upon their differentiation.

Quantum Mechanics- A branch of physics that deals with the shifting energy fields, matrices, and tiny particles that operate at or below the level of the atom. It is the microcosmic world of energy information and vibration.

Semiconductor- A substance that is capable of carrying energy as an efficient rapid flow of electrons.

Superconductivity- The flow of electrical current without the usual resistance of most metals, alloys, and other substances; it usually occurs at very low temperatures but also perhaps at room temperature in biological systems.

Tutankamen- The young pharaoh of the 18th Dynasty who returned ancient Kemet/Egypt to the worship of Amon and restored the capital to Thebes. His wealthy royal tomb was found intact in the Valley of the Kings by Howard Carter in 1922.

Vertebrate- Animals having a back bone, or spinal column. Included are fish, amphibians, reptiles, birds, and mammals.

Zygote- The fertilized egg cell before cleavage, created when sperm and egg unite.

In Memoriam: The Passing of a Giant in the Mental Health Field
T. Owens Moore, Ph.D.

Our comrade and fellow author, Richard D. King, M.D. (November 19, 1946 – December 16, 2013) made a peaceful transition to the heavenly world at his home in Los Angeles, CA. Dr. King was a major architect and contributor to the resurrection of the mind of African people worldwide. It was his work as a profound and provocative African-centered Black psychiatrist that connected him to members of the Association of Black Psychologists. Personally, his work was extremely valuable for me as a scientist searching for my place in the world scheme of history. Along with the Jegnas who taught him and the notable African historians who guided him, he blossomed into a scholar and keen researcher who walked the earth to reinterpret ancient knowledge for contemporary times.

Without his contributions, there would be a void in the mental investigation of what our ancient African ancestors gave to the world. His works on the African Origin of Biological Psychiatry and Melanin: A Key to Freedom were the written testaments he left for humanity and future generations. He was the author of the Black Dot Black Seed: The Archetype of Humanity. He was a significant contributor to the Nile Valley Conference in Atlanta, GA in 1984, and he has been the foundational presenter at many Melanin Conferences that have been held throughout the United States. It was my pleasure to meet him in 1991 in Dallas, TX at a Melanin Conference when I was a student. He was humble when I first met him, and he was respectful upon his departure from this physical world. It was an honor to move from his student to his respected colleague. In 2005, it was a delight to be a co-author on a book (Why Darkness Matters: The Power of Melanin in the Brain) with Dr. King and my fellow co-authors (Edward Bruce Bynum and Ann Brown).

King's presentations were always informative, and he was on a constant quest to liberate our minds from mental enslavement. He would often state that he was a student constantly learning. It was his personal drive and the educational foundation he received from his Jegnas at the Aquarian Spiritual Center in LA, CA and his affiliation with the Fanon Research and Development Center that made him unique in his studies.

Dr. King was always the humble servant to the community, a family-oriented man, and a trusted colleague. He made his transition to the spirit world, and he has moved on like a sun setting RA in a passing winter night. He was a magnificent scholar of African History, specifically Kemet. His studies and research have been monumental in changing our white-washed western view of history. As a scientist, he introduced different viewpoints on the topics of melanin, the pineal gland and dreams. His provocative interpretations were guiding intellectual contributions that have changed the landscape of how we as Black psychologists have redefined our reality as a formerly oppressed people.

Although he spent most of his career in Los Angeles, CA serving the community as a psychiatrist, he was well recognized beyond the west coast. In addition to knowing him as a Jegna in my life, we presented together at various melanin conferences throughout the country. He was a psychiatrist for nearly 40 years, and he expanded his knowledge beyond medicine.

In the medical field, he was priestly like a modern day Imhotep. King now resides in the etheric essence of the spiritual realm with other scholar warriors. His physical manifestation gave us so much of what we need as tools to mentally reconstruct and mold our consciousness for liberation. He understood the importance of dreams and the internal mechanisms that elevated our mental states. According to King, the African approach to dreams is the foundation of mental science, and the model of the mind was to see visions. He mastered studying the invisible realm, and many considered him a pioneer in helping us to understand the connection between the pineal gland, melatonin and melanin from an African-centered perspective.

According to our gone but not forgotten Brother, the Black Dot defines the hidden doorway to the collective unconscious; the chaos, primeval waters, universal life field that nourishes all life forms, the hidden doorway through which the transforming soul energy of Uraeus passes. Richard D. King has passed, and his soul now resides in the place he intimately investigated when he was here in his earthly form. As Black psychologists, we should recognize his contributions and may his soul rest in peace.

T. Owens Moore, Ph.D.,

Chair and Professor of Psychology

Clark Atlanta University

Atlanta, GA

AUTOBIOGRAPHICAL STATEMENTS

Edward Bruce Bynum, Ph.D., ABPP is a licensed psychologist and Diplomat in Clinical Psychology, nationally certified in biofeedback, and a senior fellow in the National Association for Applied Psychophysiology and Biofeedback. His focus areas are psychosomatic medicine, hypnosis and individual psychotherapy. He is currently in private practice in Hadley, MA.

Dr. Bynum is the author of several books in psychology and poetry. Most recent books in psychology include *Dark Light Consciousness, Our African Unconscious, The Family Unconscious and The Dreamlife of Families*. New books in poetry include The First Bird, The Magdalene Poems: Love Letters of Jesus the Christ and Mary Magdalene, The Luminous Heretic, and Gospel of the Dark Orisha.

He received the *Abraham H. Maslow Award* from APA for "an outstanding and lasting contribution to the exploration of farther reaches of the human spirit."

See: Edward Bruce Bynum, Amazon Books.com

Ann C. Brown, Ph.D., is a Professor of Anatomy and Hematology in the Biology Department at Medgar Evers College/CUNY in Brooklyn, NY. Dr. Brown teaches anatomy and physiology, general biology, chordate morphology, chordate development, and human health and disease. She has written and published articles in several journals such as *Blood* and *In Vivo*. She mentors and advises students seeking career decisions in the health sciences and medicine.

Dr. Brown developed a course for nurses called The Human Body in Health and Disease with an accompanying manual, *Laboratory Manual for the Human Body in Health and Disease*. Dr. Brown has mentored several students who participated in training programs which were sponsored by the Foundation for Advanced Education in the Sciences (FAES) at the National Institutes of Health (NIH).

Dr. Brown has presented papers at national science conferences in her areas of research as well as community groups on diseases that affect melanin-dominant populations. These areas include neuromelanin in the brain and its disease implications as well as health and food choices that impact health and well-being of African Americans. For over thirty years, Dr. Brown has been a professional health consultant. She is a member of the New York Academy of Sciences and the Metropolitan Association for College and University Biologists (MACUB).

Richard D. King, M.D., was a licensed psychiatrist and researcher in private practice in the Los Angeles County Department of Mental Health at the West Central Family Services Clinic before his transition to the ancestral realm. Dr. King was a psychiatric consultant to the Kedren Acute Psychiatric hospital, and he worked primarily with African Americans and Hispanic populations. He has presented at national conferences and was the Executor of the Aquarian Spiritual Center and a popular lecturer in the Black Gnostic Studies/Stolen Legacy series. Dr. King is the author of *The African Origin of Biological Psychiatry and Melanin: A Key to Freedom.*

Dr. King's research focus was PTSD, schizophrenia, memory recall, and stimulant and hallucinogenic substance abuse. Among his wider research interests was the exploration of the relationship of the dark matter of melanin/neuromelanin to dream states, trance dynamics and the architecture of sleep, including the neurophysiological correlates of disciplines developed to explore these states of expanded consciousness. Dr. King was a preeminent scholar in ancient Kemetic Egyptian history and psychology.

T. Owens Moore, Ph.D., has served as the Chair and Professor of Psychology in the Departments of Psychology at Clark Atlanta University in Atlanta, GA and Fayetteville State University in NC. Dr. Moore is a biomedical researcher and African-centered scholar activist involved in a wide range of interdisciplinary studies. He is trained as a physiological psychologist, and he teaches both psychology and neuroscience. His undergraduate degree is from Lincoln University in PA, and he has a M.S. and Ph.D. from Howard University in Washington, D.C.

Dr. Moore was a cofounder of the Neuroscience Institute at Morehouse School of Medicine, and he was a member of the National Science Foundation's funded Center for Behavioral Neuroscience. He is the CEO of the Melanin Institute of Hueman Technology. Dr. Moore has received federal funding to investigate the effects of hormones and neuropeptides on social behavior in rodents. He also explores learning paradigms that can be used to enhance education. He is the author of three books in this area: *The Science of Melanin; Dark Matters Dark Secrets; and Pigment Power: Topics on Melanin in Science and Health.*

See: www.drtmoore.com

NOTES

NOTES

www.ingramcontent.com/pod-product-compliance
Lightning Source LLC
Chambersburg PA
CBHW020353170426
43200CB00005B/152